CW00920928

Houses of Multiple Occupation and Selective Licensing

The aim of this book is to provide a comprehensive and up-to-date overview of the law relating to Houses of Multiple Occupation (HMOs) in part 2 of the Housing Act 2004 that local authorities environmental health practitioners use to regulate this sector. This book emerged from a set of course notes developed for a course on HMO property but took on a life of its own as the law has become increasingly complex. The law on HMOs is constantly evolving as its practical and actual application and effect on landlords is heavily dependent on the way it is interpreted by the Tribunals. This book also addresses the 2016 Housing and Planning Act, includes up-to-date key cases, and discusses how they interpret and develop the law. It also explores the limits and weaknesses of the law and the competing interpretations of key passages to illustrate where the current debates are and how they might be resolved. This book is aimed at a professional audience spread across the housing sector with enough detail to satisfy experienced environmental health and other professionals and extensive endnotes to enable further research. Other officers – housing advisers, landlords, agents – who are newer to the subject of housing law will find this a detailed introduction to the basics.

Neli Borisova is a solicitor specialising in residential property litigation. She is an associate at JMW Solicitors LLP in London. Neli advises landlords, tenants, and estate and letting agents on all aspects of residential property law.

David Smith is a solicitor specialising in residential property, agency, and regulatory law. He is a partner at JMW Solicitors LLP in London. He is well known for his work in the residential property and agency field, especially in property licensing and consumer law, and has advised local and national governments, large and small landlords, and tenants and letting and estate agents across the sector. David co-wrote the first book on the Housing Act 2004 when he was a trainee solicitor.

Routledge Focus on Environmental Health
Series Editor: Stephen Battersby, MBE PhD, FCIEH, FRSPH

Statutory Nuisance and Residential Property
Environmental Health Problems in Housing
Stephen Battersby and John Pointing

Housing, Health and Well-Being
Stephen Battersby, Véronique Ezratty and David Ormandy

Selective Licensing
The Basis for a Collaborative Approach to Addressing Health Inequalities
Paul Oatt

Assessing Public Health Needs in a Lower Middle Income Country
Sarah Ruel-Bergeron, Jimi Patel, Riksum Kazi and Charlotte Burch

Fire Safety in Residential Property
A Practical Approach for Environmental Health
Richard Lord

COVID-19
The Global Environmental Health Experience
Chris Day

Regulating the Privately Rented Housing Sector
Evidence into Practice
Edited by Jill Stewart and Russell Moffatt

Dampness in Dwellings
Causes and Effects
David Ormandy, Véronique Ezratty and Stephen Battersby

Houses of Multiple Occupation and Selective Licensing

Neli Borisova and David Smith

LONDON AND NEW YORK

First published 2023
by Routledge
4 Park Square, Milton Park, Abingdon, Oxon OX14 4RN

and by Routledge
605 Third Avenue, New York, NY 10158

Routledge is an imprint of the Taylor & Francis Group, an informa business

© 2023 Neli Borisova & David Smith

The right of Neli Borisova & David Smith to be identified as authors of this work has been asserted in accordance with sections 77 and 78 of the Copyright, Designs and Patents Act 1988.

All rights reserved. No part of this book may be reprinted or reproduced or utilised in any form or by any electronic, mechanical, or other means, now known or hereafter invented, including photocopying and recording, or in any information storage or retrieval system, without permission in writing from the publishers.

Trademark notice: Product or corporate names may be trademarks or registered trademarks, and are used only for identification and explanation without intent to infringe.

British Library Cataloguing-in-Publication Data
A catalogue record for this book is available from the British Library

ISBN: 978-1-032-28629-7 (hbk)
ISBN: 978-1-032-28639-6 (pbk)
ISBN: 978-1-003-29779-6 (ebk)

DOI: 10.1201/9781003297796

Typeset in Times New Roman
by codeMantra

Neli
To my parents Maria and Marian for always encouraging me to have opinions

David
For Aanya and Dee, always

Contents

List of Figures xi
Preface xiii
Series Preface xv

1 Introduction 1
What Is an HMO? 2
A Note on Resources and References 3

2 HMOs under the Housing Act 2004 5
The s.254 Tests 6
Interpreting the s.254 Tests 8
s.257 Test 17
Exemptions 19
Summary Position for s254 21
Potential for Change 21
Disputes 22
Some Examples 22

3 Licensing of Properties 26
Storeys 26
Mandatory HMO Licensing 28
Additional HMO Licensing 30
Selective Licensing 30
Setting Up an Additional HMO or Selective Licensing Scheme 31
Nature of Consultations 32

viii *Contents*

    Applicability of Licensing and the Duty to Licence 39
    The Application Form and the Application Fee 40
    Understanding Standards and the Licence
      Grant Process 46
    Licence Conditions 52
    The Grant and Refusal Process 57
    Liability for HMOs and Licensable
      Property and the Alternatives 58
    Companies 62
    Temporary Exemption Notices 63
    Revoking Licences and Termination 65
    Varying Licences 69
    Register of Licences 72
    Codes of Practice 74

4 Taking Over and Managing Properties   80
    Interim and Final Orders 81

5 Enforcement of the Law   82
    Prosecutions 82
    The Decision to Prosecute 82
    Time Limits 86
    Defences 86
    Mitigation 87
    Fines 88
    Civil Penalties 91
    Other Penalties 92
    Proceeds of Crime 99

6 HMO Management   104
    Application 104
    Requirements 104
    Licence Conditions 105
    Specific Obligations 105
    Active Management 108
    Risk Assessment 109
    Records 109

7	**The Tribunals**	111
	Procedural Rules 111	
8	**Conclusion**	118
	Index	121

Figures

2.1	HMO declaration (s.255) process flow	13
3.1	Licence application flow	41
3.2	Temporary Exemption Notice (s.62) process flow	64
3.3	Revocation process flow	66
3.4	Variation (s.69) process flow	70

Preface

This book emerged from a set of course notes that David had for a course on property licensing. As is often the case, the notes slowly grew as new elements were added while also becoming more comprehensive. Increasingly we were both often asked by landlords and lawyers to recommend a book on Houses of Multiple Occupation (HMOs). David had written a book on the Housing Act 2004 with Francis Davey many years ago when the 2004 Act first came into force, but this had become outdated and a simple revision was not going to cut it. Therefore, a wholly new book grew in our minds as the best solution. At the same time, the topics have grown and a book covering the entire Act would be too large a project. In addition, it is licensing and HMOs that have proven to be the most complex and dominant aspects of the Act and of the market generally.

This book is aimed at a mixed audience. It should have enough detail to satisfy legal professionals, and we have used endnotes to ensure that all of our thinking can be followed. But we have also recognised the increasing sophistication of many landlords and agents and their desire for a book that deals with the basics but also contains more detail.

Series Preface

This is the eleventh publication in this relatively new series, and more are in the pipeline. This edition is again aimed at a wider audience than usual, but reflects the wide-ranging work of environmental health practitioners (EHPs) who have to understand the complexities of the law while applying those provisions to resolving practical problems on the ground.

It illustrates the flexibility offered by the series, but the aim remains as ever: to explore environmental health topics, traditional or new, and raise sometimes contentious issues in more detail than might be found in the usual environmental health texts. It is a means whereby environmental health issues can be discussed with a wider audience in mind and perhaps prevent problems, as is the case with this edition.

This series is an important part of the professional landscape, as is apparent from the titles published so far and in the pipeline. EHPs bring their expertise to a range of situations and are deployed differently but not always to the best effect so far as public health is concerned. It may be because in some countries, governments are unaware of what is environmental health, or practitioners have a 'low profile' and are taken for granted. The work of EHPs can vary between countries, and in many cases, they do not address housing conditions, even though where one lives has considerable impact on health. It is hoped that this series will be used as a means of highlighting the work of EHPs.

At the same time, we want to encourage readers and practitioners, particularly those who might not have had work published previously, to submit proposals as we hope to be responsive to the needs of environmental and public health practitioners. I am particularly keen that this series is seen as an opportunity for first-time authors and as ever would urge students (whether at first- or second-degree level) to consider this an avenue for publishing findings from their research. Why, for example, should the hard work that has gone into a dissertation or thesis lie unread on a library shelf? We can provide advice on turning a thesis into a book. Equally this series can be a way of extending a presentation, paper, or training materials, so that these can reach a wider audience.

The series provides a route for practitioners to improve the profile of the profession as well as provide a source of information. EHPs have perhaps not been good at telling others about their work. To be considered a genuine profession and to develop professionally, EHPs on the front line need to "get published", writing up their work of protecting public health. This will allow them to analyse and report on what worked in practice, what was successful, what wasn't and why. This can provide useful insights for others working in the field and also highlight policy issues of relevance to environmental health.

Contributing to this series should not be seen merely as an exercise in gathering Continuing Professional Development hours but a useful method of reflection and an aid to career development, something that anyone who considers themselves a professional should do. I am pleased to be working with Routledge to provide this opportunity for practitioners.

As has been made clear, it is not intended that this series takes a wholly "technical" approach but provides an opportunity to consider areas of practice in a different way, for example looking at the social and political aspects of environmental health in addition to a more discursive approach on specialist areas.

Our hope remains that this is a dynamic series, providing a forum for new ideas and debate on current environmental health topics. If readers have any ideas for titles in the series, please do not be afraid to submit them to me as series editor via the e-mail address below.

"Environmental health" can be taken to mean different things in different countries around the world and so we welcome suggestions from a range of professions doing "environmental health" work or policy development. EHPs may be a key part of the public health workforce wherever they practise, but there are also many other practitioners working to safeguard public and environmental health. So, this series will enable a wider range of practitioners and others with a professional interest to access information and also to write about issues relevant to them. The format means a relatively short production time, so contents will be more immediate than in a standard textbook or reference work.

Forthcoming monographs are likely to cover such areas as Environmental Health in South Africa, Power-People-Planet (on leadership in the climate emergency), and Houses of Multiple Occupation. We are also in contact with colleagues around the world encouraging them to submit proposals. That does not mean we have no need of further suggestions, quite the contrary, so I hope readers with ideas for a monograph will get in touch via Ed.Needle@tandf.co.uk

Stephen Battersby MBE PhD, FCIEH, FRSPH
Series Editor

1 Introduction

Houses of Multiple Occupation (HMO) property is a significant area of growth in the Private Rental Sector (PRS). A 2019 Commons Library Briefing paper stated that nearly 500,000 properties were HMOs[1] and this sector is fundamental to our ability to continue to house people in the future. A 2014 study by Generation Rent suggested that over 100 Members of Parliament will have constituencies that are made up of a majority rented population by the same time.[2] Many of these people will have to be accommodated in shared housing of some sort. At the same time, the HMO sector houses some of the most vulnerable people in our society, and so it is a part of the PRS which is increasingly regulated, in part to protect those people. The number of HMOs and the number of people accommodated in them, however, continue to grow apace in response to the worsening situation around supply of reasonably priced housing in some areas, changing desires, and demographics which mean that house sharing is increasingly popular among young people, and increased financial pressures on landlords who find themselves needing to increase yield from their properties to service higher tax burdens. The changing demographic of HMO occupiers is also leading to changes in HMOs. They have started to move from being poor quality, budget accommodation to occupying a much wider space in the market with "high-end" HMOs now appearing with premium features and services.

HMOs have long been regulated to some extent both in terms of the numbers and in terms of the standard of the property and its management. However, the Housing Act 2004 radically overhauled this regulatory regime and created an entirely new structure for the regulation of these properties. As with any new legislation, it has been a long process for landlords and local authorities to understand their rights and obligations and how these intersect. This has not been assisted by the fact that the new system was not entirely well thought out and created a structure which owed more to the views of desk-bound civil servants than to the, often messy, reality of residential properties.

This complexity and confusion has spilled over into the manner in which local authorities deal with the legislation. Some of them have put in considerable effort to understand the structure and use all of their powers efficiently. Others have cherry-picked some of the easier parts out of the legislation and ignored other areas which are intended to be used with them. Some local authorities have unfortunately failed to properly understand the legislation at all and have acted in ways which the legislation simply does not permit or have behaved in an illogical and capricious manner.

At the same time, the complexity of the legislation has led to many landlords failing to properly understand their obligations and they have simply not carried out the actions the Act demands. In the worst cases, this has led to prosecution. Even worse, the uncertainty over the operation of the legislation has enabled poor quality landlords to deliberately flout the law and plead ignorance later on or use errors and misunderstandings by local authorities to avoid prosecution. This has led to a relatively low level of prosecutions under the legislation which has caused increased frustration for local authority officers and landlords who are striving to act within the law.

The situation has been further altered by the government making changes along the way. These changes have been minor at first, but more recently, a substantial set of changes have been made to the penalty regimes under the 2004 Act by the Housing and Planning Act 2016.

This text aims to explain all the key elements of the Housing Act 2004 that apply to HMOs. There is important legislation in relation to council tax and planning as well but they are beyond the remit of this book. It aims to cover the situation in both England and Wales which have similar but different regimes. It in no way deals with Scotland or Northern Ireland which merit books of their own on this topic.

What Is an HMO?

One of the difficulties associated with HMO work is the multiplicity of definitions associated with HMOs which fulfil slightly different purposes. There have been as many as four definitions in use at various times for:

1   Licensing under the Housing Act 2004;
2   Planning under the Town and Country Planning Act 1990;

3 Council tax liability under the Local Government Finance Act 1992;
4 Council tax valuation under the Local Government Finance Act 1992.

It is really important when talking about HMOs to be certain about the context in which the discussion is taking place and the specific arm of local government that the discussion is being had with. The various different definitions do not link together well and it is perfectly possible for a property to be an HMO for one purpose and not for others. The fact that a property has been approved as an HMO for one purpose has nothing to do with its approval or otherwise for other purposes and so the various definitions are all independent and should be considered separately. However, for a person dealing with HMOs as a landlord, agent, or lawyer, the different definitions are highly relevant and need to be thought about as a group as they will, in practice, interact with one another. This book only deals with point 1 above and issues involving planning or council tax are outside its scope.

A Note on Resources and References

This book refers to a number of pieces of legislation. Many of these will be to the Housing Act 2004 and regulations made to support it. Where there is a reference to a section of an Act without referencing the Act, then the reference is to the Housing Act 2004. We have tried to give a large number of references to the legislation to help guide readers through this complex web. In relation to regulations, it should be remembered that there are usually two regulations for each area, one for England and one for Wales. They are often word for word the same but this is not always the case. In general, legislation can be found on the internet through government provided websites. However, it should be noted that these may not always be up to date.

Case references are also supplied for cases in the First-tier and Upper Tribunal but not all of these cases are easily available. In particular, the First-tier Tribunal (FTT) has altered in form since its original incarnation as the Residential Property Tribunal (RPT) and that change has led to a loss of a lot of information previously found on its website, including a large number of case reports. The Welsh incarnation of the RPT had, until recently, not been publishing case reports on the internet and so remains something of a black hole in this area.

It is important to bear in mind that FTT and Welsh RPT cases are not binding on other tribunals of the same level and, while a few of them have become so commonly accepted that they are worth utilising, the majority of such decisions are specific to the facts that surround them. As time has gone on, the importance of such cases has declined as the Tribunals have matured and as a flow of decisions from the Upper Tribunal have started to appear. Accordingly, we have only referred to a very few cases from the FTT or Welsh RPT. Upper Tribunal decisions bind FTTs, but do not bind other Upper Tribunals. There have been scenarios where Upper Tribunal cases do not always appear to align well. This has been particularly notable in the area of Rent Repayment Orders. In general, later cases have tried hard not to suggest that previous cases were wrong and have usually sought to limit their scope by making them specific to the facts of that case or by suggesting that they were not giving a full statement of the law but a synopsis of it as pertaining to the current situation at hand. Due to the relatively small number of cases appealed beyond the First-tier and even smaller number appealed beyond the Upper Tribunal, this has left a number of areas of uncertainty in the law.

Notes

1 Wilson W & Cromarty H (2019) *Houses in Multiple Occupation (HMOs) England and Wales*. House of Commons Library Briefing Paper no. 0708. London: House of Commons Library. [Online] [Accessed on 21 March 2021] https://researchbriefings.files.parliament.uk/documents/SN00708/SN00708.pdf.
2 See Generation Rent, *Renter Power: Is Parliament's paradigm of home ownership coming to an end?* (London, October 2014). https://d3n8a8pro7vhmx.cloudfront.net/npto/pages/1017/attachments/original/1414665550/Renter_Power_report.pdf?1414665550.

2 HMOs under the Housing Act 2004

The creation of the Housing Act 2004 brought the topic of Houses of Multiple Occupation (HMOs) to the forefront. Prior to the 2004 Act, the definition of an HMO was rather less clear. Under the previous regime,[1] an HMO was defined by reference to households and the definition means that properties "originally constructed or subsequently adapted for occupation by a single household"[2] were an HMO. However, "household" was not defined and, in practice, each local authority largely used their own varying interpretations. This meant that there was little certainty as to how many HMOs there were and no substantial regulation. The lack of certainty also led to lax enforcement as local authority officers could not be sure of their ground. This also tended to lead to a somewhat complex system with local authorities seeking to follow guidance published by interested bodies such as Chartered Institute of Environmental Health and the case law by creating their own range of definitions of HMOs, frequently each with their own amenity standards. The 2004 Act changed this by producing a clear unified definition of a household – and from that an HMO, creating structures to enable licensing of HMOs and other property, and setting out mechanisms to provide guidance for the condition of HMOs and the manner of their management. Inevitably, a clearer regulatory regime also led to the creation of more HMOs as landlords understood what they did and did not have to licence. This was also helped along by a surge in single tenants entering the market and a general shortage of rental property at lower price points.

HMO is a phrase often used casually or without being fully thought through. Landlords will often assert that they do not have an HMO because of the way that the tenants live. It is also common to confuse the more general term HMO with a licensable HMO without realising that much of the legislation relating to HMOs applies regardless of whether or not they need licensing. This has also led to the use of phrases such as "MiniMO" to describe HMOs that are smaller and do not need licensing as if they were in some way different.

However, an HMO is defined in s.254 of the Housing Act 2004 and the definitions are reasonably clear for most purposes.

The s.254 Tests

There are in fact several tests for HMOs. There is a group of tests found in s.254 of the 2004 Act which set out what most people think of when people talk about HMOs. There is a totally separate test found in s.257 which is aimed at older properties that have been converted into flats and may have fire safety standards that fall below modern ideals.

Under s.254, the three tests are known as the:

- Standard test;
- Self-contained flat test; and
- Converted building test.

A building or part of it will be an HMO if it meets one of these tests or the separate test under s.257 of the Act (more on this later), or there is an active HMO declaration in respect of it.

The standard test is the main test from which all others spring, so it is the most important. It is also the test which applies in the majority of HMO properties. The test as set out in the 2004 Act reads[3]:

A building or a part of a building meets the standard test if—

a   it consists of one or more units of living accommodation not consisting of a self-contained flat or flats;
b   the living accommodation is occupied by persons who do not form a single household (see section 258);
c   the living accommodation is occupied by those persons as their only or main residence or they are to be treated as so occupying it (see section 259);
d   their occupation of the living accommodation constitutes the only use of that accommodation;
e   rents are payable or other consideration is to be provided in respect of at least one of those persons' occupation of the living accommodation; and
f   two or more of the households who occupy the living accommodation share one or more basic amenities or the living accommodation is lacking in one or more basic amenities.

This is not an easy read and the basic definition needs a lot of help from the surrounding legislation before it fully makes sense. All of points (a) to (f) must be met for the test to be satisfied. So, if one element is not made out, then the property is not likely to be an HMO.

The test requires a lot of unpacking and we have done so for much of the rest of this chapter.

Self-Contained Flats

The self-contained flat's test[4] applies in relation to an individual flat which meets the standard test except for the part (a) requirement not to be a self-contained flat. It is essentially the same as the standard test but applies to individual flats. At first blush, the test might seem unnecessary and the legislation might appear to have been simplified by making the standard test applicable to self-contained flats as well. However, the separation exists as it allows for different licensing treatment for flats as opposed to houses. Until recently, that differential treatment had not been utilised, but since October 2018, it has been used in England to apply different HMO licensing regimes to houses and flats.

A self-contained flat is defined[5] as a property which is physically above or below another property and in which all of the basic amenities are exclusively available for the use of its occupiers. In other words, they are not shared. The legislation is not very clear as to whether the individual flats must themselves be physically self-contained. The name "self-contained flat" implies containment but the definition is focused on the vertical location of the flat and the self-containment of the basic amenities. Therefore, it remains uncertain whether a flat which had no front door, but which had locks on internal doors such that only the occupiers could access the cooking and washing facilities would properly satisfy the self-contained flats test.

Converted Buildings

The converted building test[6] specifies that a property is an HMO if:

a   it is a converted building;
b   it contains one or more units of living accommodation that do not consist of a self-contained flat or flats (whether or not it also contains any such flat or flats);

c   the living accommodation is occupied by persons who do not form a single household (see section 258);
d   the living accommodation is occupied by those persons as their only or main residence or they are to be treated as so occupying it (see section 259);
e   their occupation of the living accommodation constitutes the only use of that accommodation; and
f   rents are payable or other consideration is to be provided in respect of at least one of those persons' occupation of the living accommodation.

The phrase "converted building" is defined as meaning "a building or part of a building consisting of living accommodation in which one or more units of such accommodation have been created since the building or part was constructed".[7]

So, the converted building test is essentially a hybrid test intended to catch properties which have been adapted after construction to contain a mix of flats and individual rooms with shared facilities. This test will apply to buildings which contain some self-contained flats and some accommodation which share basic amenities. These properties will be HMOs under this test even though parts of them will not share any basic amenities. However, it is wider than this and will pick up any property that has been converted such that more units of living accommodation have been created since original construction and there are shared basic amenities.

The test will apply regardless of what the property was before, provided there has been adaption. The date of the adaption, who carried it out, or even the fact that the current owner is unaware of the adaption are all irrelevant. So, some owners of properties that they may believe are self-contained flats but which were in fact single town houses several hundred years ago have been caught out. The test will also catch former commercial property converted to residential accommodation where there is sharing of basic amenities.

Interpreting the s.254 Tests

There are aspects of the s.254 test which cause confusion and are often misinterpreted. In addition, there are parts of the test which are poorly drafted and have multiple competing interpretations. Some of these issues have been resolved by the courts and Tribunals, while others have yet to be dealt with.

Buildings and Parts of Buildings

The tests are all clear that they involve a "building or a part of a building". The second element of this phrase is often ignored. In practice, it can be possible for a local authority to bypass some of the problems with the HMO definitions (especially the sole use test) by simply choosing to scope the HMO more narrowly around a specific part of the building. In addition, it would theoretically be possible for a local authority to avoid suggesting that any specific property is an HMO by simply scoping the parts of the building very narrowly to talk about one dwelling at a time. Of course, a local authority like any public body must make these decisions in a reasonable and rational manner and so more extreme examples of this are unlikely to occur in practice.

Units of Living Accommodation

There are a range of complex problems with the definition of an HMO. The first springs from the somewhat lazy use of language around the phrase "unit of living accommodation". The standard test uses the phrase "units of living accommodation" as a part of the test, but this is never defined. Other parts of the test use the phrase living accommodation in a much more general fashion. The problem with this is that the concept of a "unit of living accommodation" is then left somewhat open to interpretation.

Consider a property with five bedrooms and shared kitchen and bathroom. Is the entire property a unit of living accommodation because a unit of living accommodation surely needs a kitchen and bathroom? Alternatively, is each bedroom a single unit with the kitchen and bathroom being adjunct rooms? The standard test actually uses living accommodation in both senses within it, so the point is not clear. The interpretation thus far has tended toward the idea that each bedroom is one unit as otherwise the legislation would be entirely useless but the point has never been properly settled. A more practical question would probably be how to deal with hybrids in which some rooms had bathrooms while others had bathrooms and kitchens.

Only or Main Residence

Another area of uncertainty relates to part (c) of the standard test. That is the requirement to use the property as the occupier's only or main residence. Most scenarios are easily resolved as the persons are clearly occupying a property as their main residence. The 2004 Act identifies specific scenarios which are always seen as a main residence.[8] These are students in full-time education and people using the property as a refuge. However, none of this resolves the issue of mixtures of people. Not every person may be a full-time student, there may be part-time students or non-students mixed in with them. In addition, the 2004 Act has no definition of what constitutes a full-time student and there is no agreed definition as different areas of legislation define full-time education using varying numbers of teaching hours.

Part (c) is unclear about these different scenarios. When it says that persons are occupying as their main residence does that mean if some of them are then they should all be seen as doing so? Should we simply not count those who do not occupy as their main residence, or must everyone occupy as a main residence in order to trigger that element of the test?

In addition, the Act allows for further categories of persons to be added who are also considered to be occupying as their only or main residence.[9] The relevant regulations have added migrant workers, seasonal workers, and asylum seekers to this group.[10] The definition of a migrant worker was altered in England as a result of the UK's departure from the European Union (EU).[11] It now specifies that a migrant worker is any person who has obtained permission to enter or remain in the UK by way of a residence scheme and has taken up employment or any person who has been permitted entry for the purpose of taking up employment. This is actually a more logical definition in that it is more balanced. The original definition of a migrant worker only applied to European Economic Area (EEA) nationals who were residing in the UK under EU free movement rights and so excluded non-EEA nationals who were working here temporarily. So, under the old regulations a Hungarian construction worker here on a temporary project for a few months would be a migrant worker but one from neighbouring Serbia would not be. The new structure is simpler and would mean that both groups were treated as occupying the property as their home.

A seasonal worker is a worker whose employment varies according to the seasons but repeats every year and that period of employment does not exceed eight months. In both these cases, the accommodation must be being provided by their employer or an agent of their employer in order for them to be classified as occupying as their only or main residence. If they are occupying accommodation of their own volition, then they are not assumed to be occupying as their main residence, although this can still be demonstrated on a case-by-case basis. The definition of an asylum seeker is drawn from other legislation[12] but provided an occupier meets that test then they will be deemed to occupying as their only or main residence regardless of who is housing them.

Sole Use

Similar problems arise in relation to what is usually called the "sole use" element in part (d) of the standard test. This states that the occupiers' use of the living accommodation must constitute "their only use of that accommodation". S.260 tells us that there is a presumption that the sole use condition is met unless the contrary is shown. However, it is not clear what sole use means. If a property has a mixture of HMO tenants and other rooms let as a guesthouse or B&B does that mean the sole use condition is not met? Or is it met in relation to the parts of the property being used as an HMO but not in relation to those parts being used as a B&B?

As an alternative viewpoint, one of the authors has had it from government lawyers that the sole use condition is intended to refer to the perspective of the occupiers and how they use the premises. This might be indicated by the reference in the condition to occupiers' use of the living accommodation being *their* only use. "Their" in that interpretation meaning the use made of the accommodation by the occupiers. So, in that case, the meaning of "sole use" is that the occupiers must be solely using the premises for residential accommodation and not for a mixture of residential and business use. In other words, the focus is on how the occupiers themselves use the building and not on how the owner is using it. So, if the tenants have rented a property from a landlord on a lease which permits the tenants to use the property as a mixture of residential and commercial space, then the sole use condition would not be met. But even this statement is unclear. What if I have rented a shop with accommodation above it from a landlord as part of a single tenancy? Presumably, this means that the sole use condition would not be met and so the property would not be an HMO from that landlord's perspective. However, what if I then rent the accommodation out separately from the shop? Would the accommodation I have rented out then meet the test for sole use as the ultimate tenants of the accommodation would be using solely as living accommodation, even though my use of the entire demise is mixed.

The legislation is entirely unclear as to whether the reference point is the use of the landlord, a tenant, or the end occupiers.

The Upper Tribunal (UT) has weighed in on this issue, at least in relation to property guardian scenarios.[13] First, the UT has refused to accept any equivalency between the phrase "sole use" as set out in the Act and the phrase "sole purpose". In relation to property guardians, the argument being deployed was that the building they were guarding was primarily commercial premises. So, the "purpose" of the premises was as commercial premises and there was also a purpose in providing accommodation for the guardians. Therefore, there were two purposes and so residential accommodation was not the "sole purpose". The UT did not accept this argument and was clear that "use" meant just that. As the premises in question were not being used for commercial reasons, the "sole use" was residential accommodation and the test was satisfied. In summary, the UT seems to be adopting neither of the viewpoints laid out above and is taking an objective but flexible approach to the situation looking at how the accommodation is being used on the ground.

HMO Declarations

There is another means for local authorities to deal with the sole use problem, at least in part. That is through the use of an HMO declaration (Figure 2.1).[14]

The Act allows a local authority to take the view that a property, or part of it, is an HMO, even if the sole use condition has not been met. If all the other parts of one of the HMO tests are met and the HMO use is a significant use (rather than the sole use), then the local authority has a power to make a declaration that the property is in fact an HMO. It is possible for the landlord to challenge that designation by way of an application to the First-tier Tribunal (FTT) within 28 days.

The difficulty with this system is that many local authorities tend to assume that properties are HMOs without stopping to consider whether an HMO declaration should be made. Therefore, there are not many useful appeals to the FTT to consider.

However, in the relatively few cases that have come before it, the FTT has been generous to the local authority and has tended to uphold declarations. In part, this is because many of the appeals against such declarations have been misconceived but also because the applicant has not taken proper account of the specific presumption in the Act that the sole use condition has been met and the further presumption that the significant use condition has been met.[15] In most cases the applicant has generally not done enough, and has often presented very weak evidence, to overturn the presumption in favour of the significant use condition. Anyone seeking to challenge such a declaration should ensure that they have very clear evidence of a mixed use that means the HMO use is wholly insignificant. It is also worth noting that

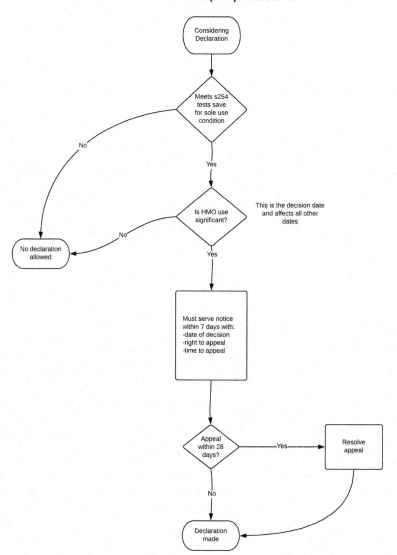

Figure 2.1 HMO declaration (s.255) process flow.

insignificant use in legal terms essentially means almost no use at all so even the smallest amount of HMO use is likely to be enough to meet the significant use test.

It should also be noted that it is not open to a landlord to simply clear a property of people and then seek to immediately overturn an HMO declaration where the local authority have clear evidence that the property is an HMO. The UT has warned that this must be guarded against and Tribunals will weigh up the evidence of the local authority and landlord where there is a dispute.[16] If a property has stopped being used as an HMO, then a request can be made to revoke an HMO declaration (see below) and the local authority should consider that request and make the appropriate decision.

What Is a Household?

For a property to be an HMO, it must consist of more than one household.[17] The definition is somewhat counter-intuitive as it has nothing at all to do with how people live and is based on their personal relationships. To that end, it is focused around families and cohabiting couples. Therefore, persons will be regarded as forming a single household where they are a couple or where they are related as:

- Parents;
- Children;
- Siblings;
- Uncles and aunts;
- Nephews and nieces;
- Grandparents;
- Grandchildren; or
- First cousins.

Foster, half, and step relationships all count as full relationships and so a stepchild is still a permitted relationship for the forming of one household. However, in-laws are not in the list and they will not form part of the household. The list of relationships that form a household may be added to in secondary legislation and certain personal employees of the tenant were added in this way after the Act was brought into force.[18] In principle, there is no reason why further groups could not be added in future.

It is important to be clear that the definition of "household" used in the legislation is not one that most ordinary people, or an English dictionary, would recognise. A household is created by a group of people with the defined familial relationships set out above. If there is no such defined relationship, then they will be separate households. So, there is nothing to stop two individuals in a property from being two households. The age of the occupiers is totally irrelevant. The 2004 Act is entirely unconcerned with the manner in which people live when considering whether they form a household and is solely focused on the family relationship. Therefore, three friends sharing together would form three households while three cousins would form one household. This would be true even if the three friends shared their bills through a joint account and cooked their meals collectively together, because they lack the necessary bond of blood to form a household for the purposes of the Act. By the same token, the three cousins would form one household even if they did not share bill payments or cooking and indeed if they never spoke at all, because they would have the necessary legally defined relationship. While this definition is slightly counter-intuitive and takes little account of the actual situation in each home, it has the advantage of being very clear in its operation. However, it does create a difficulty for enforcement purposes. Parental relationships are easy to show using birth records but relationships such as uncles, aunts, and cousins are rather harder to demonstrate. It is also the case that these appellations are used more casually in some parts of British society and can be applied to people with no true blood relationship at all. The inclusion of cousins, in particular, leaves both landlords and enforcers vulnerable to false statements on the part of tenants who might assert that they are cousins in order to secure a property even where this is untrue and it might be difficult for anyone to demonstrate that this is untrue, especially if they come from another country with a different system of recording births and relationships.

The household definition also suffers from operating as a network of relationships. This is because there is no central measurement point. Therefore, when considering the relationships involved in creating a household, it should be thought of as more of a chain than as a group with a leader. There is no "head of the household" or nuclear family in the classical sense. This means that an occupier along with their cousin, and their cousin's cousin, will still all be seen one household. In principle, this means that two people who have the most tenuous relationship could still be one household provided that there are other occupiers who act to form a chain of acceptable relationships between them. More practically, a household consisting of an individual living with their spouse and the spouse's sister will still be one household for the purpose of the 2004 Act because a chain of relationship from the individual through to the spouse's sister will be maintained through the spouse.

The definition of "household" is also limited to cohabiting *couples* and does not allow for different arrangements, such as a polyamorous throuple for example.[19] This will presumably have to be addressed in secondary legislation at some stage if the definition is to keep up with societal change.

The regulation also brings within the definition of household a list of domestic staff. This includes, but is not limited to:

- Au pairs;
- Carers;
- Governesses;
- Chauffeurs;
- Secretaries;
- Personal assistants;
- Gardeners; and
- Servants.

This only applies where the staff member is working exclusively in a domestic capacity for the occupier of the property or any part of it or for any member of the occupier's family. In addition, the staff member must not be paying rent, other money, or anything else for the accommodation which must be provided solely as part of the compensation for their work. There are difficulties with these requirements in that there is no definition of what constitutes the occupier's family for the purpose of employment of a member of domestic staff. In practice, this probably means falling within the range of approved relationships that make up a "household" and being resident in the same property. The definition is also open to a degree of interpretation as there is potentially a wide range of domestic roles that will not fall into the suggested list set out above. As a guideline, the staff member must be working full time for the occupiers and working for them rather than for any business that they are associated with. In the vast majority of situations, it is probably the "au pair" that is likely to be the domestic staff member that is relevant for the household definition.

In addition, where a person is part of the tenant's household because they are domestic staff, then the family of that domestic staff member will also be a part of that household, so an au pair who also had her own child would find that both would be considered part of the tenant's household.

As well as domestic staff, regulations have specified that certain carers are also part of the occupier's household.[20] This applies as long as the carer is an approved adult placement carer and as long as that carer is not caring for more than three persons at that place.[21] Where a person is living with a foster parent, the requirement is that they have been placed with that foster parent.[22]

Tenants, Licensees, and Numbers

It is important when considering a property to remember that the nature of the occupation agreement is not important, simply that a person is occupying a property as their home for money. Therefore, the fact that a person is not a tenant but is instead a licensee is not relevant. Some landlords have sought to get around the legislation by granting licences instead of tenancies, not naming some occupiers as tenants on written agreements, or letting to a company which then allows employees to occupy the property. None of these devices will work as the nature of the relationship between the occupier and the person allowing them to occupy is simply irrelevant. The fact of occupation by a group of individuals is sufficient. A similar issue applies when considering the number of occupiers. This cannot be evaded or cheated by unusual constructions. The issue is whether or not an individual occupies a property as their principal home. If they do, then no amount of adjustment to the manner in which the relationship is described is relevant. By the same token, the age of the occupier is not relevant. A mother with a baby that has just been born will count as two occupiers just as much as a mother and 15-year-old daughter. They will both be one household because of the relationship between them but there will be two occupiers. While there is a power in the Act[23] for secondary legislation to be made which discounts occupiers of a certain age or to count them as fractions of a total occupiers, this power has never been utilised. This creates a trap in some cases, in that once the threshold is reached then every occupier will count. So, for example, if I am living with my wife and two lodgers in a three-storey house, then there are four of us in the property in three households (myself and my wife as one household and each lodger as a household of their own), but it is not an HMO as the two lodgers are ignored (see the section on exemptions below). If, however, there are three lodgers, then the exemption is no longer met and the total number of occupiers is five (three lodger plus myself and my wife). Accordingly, the property will need a licence.

s.257 Test

There is an entirely separate definition of an HMO under s.257 of the Act. This states that a whole building is an HMO for the purposes of that section where:

- It is a building which was a single dwelling but has been converted into multiple dwellings;

- The conversion does not accord with the Building Regulations 1991 or if constructed later is not in accordance with the building regulations relevant to that period and still does not meet that standard;
- One third or more of the properties are not owner-occupied.

In this context, owner-occupied means that the flat is occupied by a person who owns the freehold of the building, a person who has a lease of the flat which was originally granted for more than 21 years (that is the original lease term, not the residue of the lease), or a member of their families.

The fact that a building is an HMO under s.257 is totally unconnected with the occupation of the individual flats and whether or not those are HMOs under s.254. In other words, a building could be a s.257 HMO and one or more of the flats within it could also be s.254 HMOs depending on their occupancy. In fact, the legislation is clear that the fact that a building as a whole is an s.257 HMO has no effect on the flats or other accommodation within it and their HMO status is unaffected.[24]

It should be noted that if the building has been improved over the years by upgrading it such that it does now comply with the Building Regulations 1991 then the s.257 test will not apply. It is well worth making sure that if this work is done, a report is obtained from a chartered surveyor or other suitably qualified expert confirming that the property does now comply with the regulations. Anecdotal evidence suggests that local authorities will not accept statements that the work has been done and will even pursue a prosecution unless very clear evidence is produced that the property now meets the standards.

In practice, it is quite hard to demonstrate whether a property does not comply with the Building Regulations 1991. The only way to be sure is to start looking at the inside of the walls and ceilings. Local authorities are empowered to access premises to carry out a survey to decide whether they should exercise any of their powers under other parts of the Housing Act 2004.[25] This includes a survey to determine whether or not the property needs a licence. Under the same powers of entry, an officer is entitled to take measurements and recordings or to leave equipment for recording and later collection. However, there is something of a flaw in this legislation. While it empowers officers to take samples for testing of any "article or substance found on the premises", it is unclear whether this would extend to taking samples of the premises itself.[26] This creates a problem for determining whether a property meets the Building Regulations 1991 standard or not as it would be almost impossible to determine this without taking a sample of the property concerned. In practice, local authority officers often assume on quite limited evidence that a property does not meet the relevant standard, sometimes with approximations or using a weak reading of the relevant Building Regulations standards. However, this approach ignores the fact that ultimately, for a local authority to take formal action, they are going to have to be able to prove that a property meets the s.257 test to the criminal standard of proof. That means that the evidence will have to be substantial and supposition of the type that is often advanced is unlikely to be sufficient.

A somewhat peculiar situation was considered by the UT in 2020 where a long leaseholder (who also co-owned the building) sought to challenge a finding that the building was an HMO under s.257 of the Act.[27] The Council's finding was based however on an HMO licence application submitted by another leaseholder in which they had stated that the building did not comply with the 1991 Building Regulations.[28] The Respondent challenged that statement but then provided no evidence of compliance with the 1991 regulations. Given that she was seeking to impugn a statement made by another building owner it was in that case for her to prove compliance with the 1991 Regulations. Some local authorities have, however, misinterpreted this decision as meaning that they do not have to prove failure to comply with the 1991 Building Regulations and it is for the owner to demonstrate the fact of compliance to them. That is not the case and this decision is fact specific. If a local authority is seeking to take action for a lack of 1991 regulations compliance and an assumed s.257 HMO status, they will need to prove their case. In practice, this may make it very hard indeed for a local authority to pursue a prosecution for failure to licence under s.257.

Exemptions

There are some key exemptions which, if met, prevent a building or part of it being an HMO under either s.254 or s.257 of the Act.[29]

The most important exemption for most people is that a property occupied by two persons in two households can never be considered an HMO. This is sometimes missed by inexperienced local authority officers. So, two friends sharing is never an HMO. Three friends sharing will be one.

There are also exemptions for properties controlled entirely by various organisations. These include those where the person managing or having control is:

- Public bodies including local authorities, non-profit registered social housing providers and social landlords, profit making social housing providers where the property is social housing, fire authorities, policing and crime commissioners, and health services;
- Educational institutions as long as they are within an approved list set out in the relevant regulations.[30] The building must be occupied solely or mainly by students who are doing so in order to undertake full-time education; and
- Religious communities where their principal occupation is prayer, contemplation, education, or the relief of suffering. However, the exemption for religious groups is not applicable to s.257 HMOs.

Properties managed or controlled by these groups are not HMOs.

Where a property is managed or controlled by a housing cooperative society, it is also exempt. However, the cooperative society must be registered and the membership must be restricted to those persons who occupy or intend to occupy the property concerned and those members must have direct control over its operations on an equal vote basis. None of the occupiers can be occupying the property under an assured (or assured shorthold),[31] secure,[32] or protected[33] tenancy. In other words, these properties are only exempt if it is a genuine cooperative society with the members being occupiers and in genuine control over the premises as opposed to a cooperative set up by a controlling party who in practice controls the vote share and where other occupiers are members without any true powers.

Where the owner also lives in the property, it is also exempt to a certain extent. This exemption only applies to s.254 HMOs. Owner in this case means a person who owns the freehold or has a long lease of more than 21 years as well as the members of their household. In this case, the definition of household is as for other s.254 HMOs and so will only apply to immediate family members. The number of additional persons outside the owner and their household which is permitted before the property becomes an HMO is set by separate regulations. At the moment, the number is set at two.[34] This means that an owner can reside in their property with any number of members of their own household and anything up to two additional persons who are paying rent and are not part of their household without the property being an HMO. However, if the number of additional persons reaches three, then the property will be an HMO. This exemption is often breached and it is not uncommon to find properties with owners living with a substantial number of lodgers, making the property an HMO. As older people seek to remain in larger properties and younger people try to get on the property ladder, this problem appears to be growing.

Another exemption is associated with properties regulated under alternative regimes. This is intended to exclude care homes of various sorts and prisons from the ambit of HMO licensing by specifying that they cannot be HMOs. The exemption is created due to the specific legislation that the property is regulated under.[35] However, not all pieces of legislation are covered. So, for example, it is commonly thought that being regulated by the Care Quality Commission (CQC) provides an exemption. But the regulatory regime created by the CQC is defined by the Health and Social Care Act 2008 which is not an Act set out in the relevant regulation and so merely being regulated by the CQC is not enough to avoid HMO status on its own.

Summary Position for s254

In summary, the above means any property (including flats) or any part of it:

- Which has three or more occupiers;
- Where those occupiers form more than one household;
- Where rent or its equivalent is being paid;
- In which at least one of the occupiers lives in the property as his or her principal home; and
- Where at least two of the households share basic amenities, such as cooking or washing facilities.

Is likely to be considered an HMO for the purposes of s254.

Potential for Change

The HMO definitions can be changed by statutory instrument. In other words, no formal Act of Parliament is needed and so they can be changed relatively easily by way of an order from the Secretary of State or the Welsh Ministers as appropriate. However, there appears to be little desire to tamper with the s254 definition at the current time. Even if there were a desire to make changes to the definition, then any new definition would probably be just as complex and would likely end up having its own range of issues and flaws. There is always a balance between complexity of the definition chosen which then limits the scope of the legislation to properties of concern and creates the risk of confusion and loopholes as against a simpler definition which can end up being overly broad and create regulation of properties that do not require it.

Disputes

There is a difficulty when a dispute occurs over whether or not a property is an HMO. There is no power for the Tribunal to settle disputes over whether or not a property is an HMO. The only option is for the local authority to commence a prosecution or seek a civil penalty and then let the Tribunal or the courts resolve the issue. However, this means that for a landlord the consequences of such a dispute can be severe as if they are wrong they will be fined or prosecuted. Given the complexity of the legislation, it would have been sensible to include some form of lower risk mechanism to deal with such disputes but Parliament chose not to do so. In a few cases, local authorities have worked cooperatively with landlords where there is a clear uncertainty in the law but in general this has not happened and these issues have been resolved on the back of prosecutions or civil penalties.

Some Examples

The different HMO definitions and the complex interplay of different provisions lead to a great deal of confusion over whether or not individual properties are HMOs. This causes difficulty for local authorities as much as for individual landlords. It is often helpful to consider some examples of different scenarios.

1. The property is a two-storey house occupied by a married couple, their daughter, and the cousin of the husband.

    This is not an HMO because the occupiers form one single household. The cousin is related to the husband and he is related to everyone else. If the husband were to leave the home, then it would become an HMO as the cousin would no longer be related to the other two occupiers within the meaning of the regulations.
2. The property is a large family home which is rented to a single limited company. This company uses the home to accommodate five of its employees who are not related.

    This property is an HMO. There is only one tenant, the company, but there are five occupiers who do not form one household and they share basic amenities. Additionally, the employees would be deemed to be occupying as their main home if they were migrant workers. It would only be possible to assert that the property was not an HMO if the occupiers were all UK nationals and it could be shown that they were not occupying the premises as their home.

3 The property is a purpose-built block of flats containing five individual flats. The block is owned by a limited company and the five flat leaseholders each own a share of the company and are directors. Three of the five flats are sub-let on Assured Shorthold Tenancies to tenants who have been there between one and three years. Two of these are families while one has three unrelated students sharing in it. The remaining two flats are occupied by the long leaseholders.

The block of flats is not an HMO as it is purpose-built and so does not qualify for the s257 HMO test which only applies to converted properties. The flat with the three students in it is an HMO as it is let to three unrelated people who share basic amenities.

4 The property is a large house that has been converted into five flats. All the conversion work complies with the Building Regulations 1991. Two of the flats are more in the way of bedsits and are occupied by an elderly lady and an elderly couple on Rent Act 1977 tenancies. These properties have their own kitchenettes but share a bathroom and separate toilet between them. The remaining three flats have been fully updated into modern self-contained studios and are all occupied by individuals.

The property as a whole is an HMO under the converted building test in section 254. Some of the flats are self-contained, but two are not and share a basic amenity. Therefore, the entire building falls to be considered. There are six occupiers who live in five households (the elderly couple makes one household while the remaining occupiers form the other four). Rent is being paid and it is their principal home.

5 The property is an old house converted into four flats in the mid-1980s, little work has been done on the structure since. All the flats are let on long-leases of 90 years with 18 years left to run. One of the owners of a lease owns the entire property and resides in her flat. Of the other three owners, two reside in their flats while one has now moved out and has sub-let his flat to a young couple on a two-year tenancy.

This property is not an HMO at the moment. It is converted into flats and does not comply with the Building Regulations 1991. However, at the moment, only one of the flats is occupied on a lease of less than 21 years, as the other flats were all let on 90 year leases, even though there is less than 21 years remaining on them. Therefore, only a quarter of the flats are occupied on a lease of less than 21 years. If another leaseholder was to move out and sub-let their flat for one year, then the property would become an HMO as at that stage half of the flats would be let on leases of less than 21 years.

Notes

1. Provided by Part XI, Housing Act 1985.
2. s345, Housing Act 1985.
3. s.254(2), Housing Act 2004.
4. See s.254(3), Housing Act 2004.
5. In s.254(8), Housing Act 2004.
6. In s.254(4), Housing Act 2004.
7. See s.254(8), Housing Act 2004.
8. In s.259, Housing Act 2004.
9. S.259(2)(c), Housing Act 2004.
10. By way of regulation 5 of the Licensing and Management of Houses in Multiple Occupation and Other Houses (Miscellaneous Provisions) (England) Regulations 2006 and the Licensing and Management of Houses in Multiple Occupation and Other Houses (Miscellaneous Provisions) (Wales) Regulations 2006.
11. By regulation 67 of the Immigration and Social Security Co-ordination (EU Withdrawal) Act 2020 (Consequential, Saving, Transitional and Transitory Provisions) (EU Exit) Regulations 2020.
12. See s.94, Immigration and Asylum Act 1999.
13. *Global 100 v Jimenez & Ors* [2022] UKUT 50 (LC).
14. Under s255, Housing Act 2004.
15. See s260, Housing Act 2004.
16. *Herefordshire Council v Rohde* [2016] UKUT 039 (LC).
17. See the definition in s258, Housing Act 2004.
18. By regulation 3 in the Licensing and Management of Houses in Multiple Occupation and Other Houses (Miscellaneous Provisions) (England) Regulations 2006 and in the Licensing and Management of Houses in Multiple Occupation and Other Houses (Miscellaneous Provisions) (Wales) Regulations 2006.
19. See the Property Ombudsman case study of 23 August 2021 available at https://www.tpos.co.uk/news-media-and-press-releases/case-studies/item/left-out-of-legislation-hmos-and-a-household-of-three [accessed 6 March 2022].
20. In regulation 4 of the Licensing and Management of Houses in Multiple Occupation and Other Houses (Miscellaneous Provisions) (England) Regulations 2006 and in the Licensing and Management of Houses in Multiple Occupation and Other Houses (Miscellaneous Provisions) (Wales) Regulations 2006.
21. As approved under the Adult Placement Schemes (England) Regulations 2004 or the Adult Placement Schemes (Wales) Regulations 2004 for Wales.

22 Under the Fostering Services Regulations 2002 or the Fostering Services (Wales) Regulations 2003.
23 Section 264, Housing Act 2004.
24 Section 257(5), Housing Act 2004.
25 See section 239.
26 See s239(8).
27 *Hastings Borough Council v Linda Turner* [2020] UKUT 0184 (LC).
28 Building Regulations 1991 (S.I. 1991/2768).
29 Set out in Schedule 14, Housing Act 2004.
30 The Houses in Multiple Occupation (Specified Educational Establishments) (England) Regulations 2013 or the Houses in Multiple Occupation (Specified Educational Establishments) (Wales) Regulations 2006.
31 Housing Act 1988.
32 Housing Act 1985.
33 Rent Act 1977.
34 See regulation 6(2) of the Licensing and Management of Houses in Multiple Occupation and Other Houses (Miscellaneous Provisions) (England) Regulations 2006 or the Licensing and Management of Houses in Multiple Occupation and Other Houses (Miscellaneous Provisions) (Wales) Regulations 2006.
35 Set out in Schedule 1 of the Licensing and Management of Houses in Multiple Occupation and Other Houses (Miscellaneous Provisions) (England) Regulations 2006 or the Licensing and Management of Houses in Multiple Occupation and Other Houses (Miscellaneous Provisions) (Wales) Regulations 2006.

3 Licensing of Properties

Not every property needs a licence. Some properties can be Houses of Multiple Occupation (HMOs) without needing licensing and some properties will not be HMOs at all. For many landlords, this remains an area of confusion with many of them thinking that their property is not an HMO because it does not fall within the definition of a licensable HMO. This chapter considers the need for licensing as well as some of the key issues surrounding it.

Storeys

Before considering licensing, it is important to consider how the legislation looks at the number of storeys in a property. Different local authorities will apply licensing requirements to different property sizes and this is usually calculated on the basis of storeys.

When calculating the number of storeys in a property, it is necessary to include basements or attics which are converted or adapted for use as part of the living accommodation (as opposed to merely used for storage) or are being used in that manner, mezzanines which are more than mere access to another level, any basement level which includes the main entrance to the property, or any business premises above or below the property irrespective of whether they are connected in any way.

In Wales, there can be some confusion about this as well. This is because when considering storeys above and below the main property, the relevant Welsh regulations[1] are very unclear as to whether one should only consider business premises or whether other types of use should also be counted. The relevant part of the Welsh Order reads:

DOI: 10.1201/9781003297796-3

c   where the living accommodation is situated in a part of a building above business premises, each storey comprising the business premises;
   ...
f   any other storey that is used wholly or partly as living accommodation or in connection with, and as an integral part of, the HMO.

The High Court has resolved parts of this issue[2] when considering a block in London. The block contained several storeys each of which comprised one self-contained flat as well as a single ground floor level which contained business premises. These were all owned by the same company. The local authority at the time only licensed HMOs which extended over three storeys so on the face of it the flats were not licensable. However, the local authority took the view that other residential storeys in the same block which contained other flats owned by the same company should also be counted as the Order stated that other storeys used as living accommodation should also be counted. Therefore, each flat, while only individually physically extending to one storey, should be seen as consisting of several storeys by way of the definition used in the legislation and therefore every flat in the block had to be licensed.

The High Court disagreed with this analysis and that the regulations do not intend that storeys used for residential purposes which are not part of the actual property being considered should be counted even though they are in the same building. Therefore, when counting the number of storeys, there is only a need to count other storeys outside the main property being considered when they are business premises. However, where a business premises is present above or below, it must be counted as a storey included within the HMO irrespective of whether it is in separate ownership or even if it is several storeys removed from the residential premises.

Stairwells and access spaces are also not counted. So, if a level is the entry to a flat, even if it crosses another level to get there, then those additional entry levels are not counted. Therefore, a two-storey maisonette on levels two and three with a private entry stairwell leading up from the ground floor is still only a two-storey property. This is the case even if there are landings on the stairwell provided they are only a turn in the stairs. However, if that entry stairwell has anything more in it than access, then that would be an additional storey and would need to be counted. So, a stairwell with a landing which had a toilet off it or a larger landing containing a desk with study space would both lead to the landing being counted as an additional storey.[3]

However, this still leaves considerable uncertainty. The regulations still say that attics which are used "in connection with, and as an integral part of, the HMO" count as separate storeys. However, an attic is not clearly defined. As a matter of plain English, an attic is a room below a roof. It is probably different from a loft but that still leaves considerable scope for uncertainty. In addition, if an attic exists but is being used merely as ad hoc storage then is it being used "in connection with" the HMO? Arguably if the tenants use it then it is. However, what if the tenants only use it a little, while the landlord also uses the same space. Is it then being used as part of the HMO and is the use of the space still exclusive use in which case does that space form part of the HMO at all?

In England, the position is more uncertain still as the English version of the Welsh Order has been repealed along with much of its definitional information regarding what is and is not a storey. This was done when England moved to licensing all properties with five or more occupiers regardless of the number of storeys. But it means that local authority schemes which licence properties with different numbers of occupiers based on their size often lack any clear basis for calculating the number of storeys as they purport to rely on the repealed English Order.

Mandatory HMO Licensing

Not all HMOs require licensing. Individual local housing authorities have a degree of discretion as to what HMOs they wish to licence. However, some larger s.254 HMOs have been decreed by the relevant governments as needing a licence throughout England and Wales regardless of the views of the individual local authority. Those properties which meet the mandatory licensing definition will always require licensing. All other types of HMO will only require licensing if the local authority has introduced an appropriate licensing scheme which deals with that type of property.

At the time the HMO legislation first came into force, the definition of an HMO subject to mandatory licensing[4] was a property which:

1   Is an HMO under the s.254 test;
2   Is comprised of three or more storeys; and
3   Has five or more occupiers.

All three of these had to be true but if all three elements were met then a licence must be obtained. From 1 October 2018, the definition of properties which require mandatory licensing changed in England.[5]

Wales continues to use a definition which is unchanged from the original English definition.[6] In Wales, all three of the conditions must be fulfilled. So, neither of a two-storey property which has five occupiers or a three-storey property with four occupiers would meet the requirement for mandatory licensing in Wales.

The amended mandatory licensing definition for England applies to properties, both houses and some flats, which are:

1 HMOs under the s.254 test; and
2 Which are occupied by five or more persons.

Any house which meets the above test will require licensing. For flats, the position is more complex. If the flat has been created by conversion from a larger single dwelling, then each individual flat that meets the above test will require licensing. If the flat is part of a purpose-built block of flats, then it will not require licensing, even if it meets the above test unless it is part of a purpose-built block that only contains two flats. This is probably a rare occurrence but it is worth bearing in mind that a purpose-built shop with two flats above it would require licensing if it were occupied within the above definition. Naturally, flats that are not picked up by this definition may be picked up by a local authority additional licensing scheme (see below). In this context, we are talking about the individual flats within a block as opposed to the entire block. So, for a flat to be a licensable HMO for mandatory licensing, it must have five or more occupiers who do not form one household and who share basic amenities. If a flat only has four occupiers, then it will not meet the requirement to be licensed under the mandatory definition, notwithstanding the fact that other flats in the same building might also have other occupiers who if counted as well make a total greater than five.

Properties that meet the converted buildings test will also need licensing if they have five or more occupiers.

This means that in England the number of storeys became irrelevant for the purpose of mandatory licensing. Landlords who had sought to avoid licensing by only having properties which were two storeys would now be caught if they had five occupiers. Some landlords no doubt altered their model to limit themselves to four occupiers in order to stay outside the amended licensing definition. This is entirely permissible but a landlord doing this will, of course, be limiting the revenue that they will recover from the property so the benefit in avoiding licensing is quite small.

Additional HMO Licensing

Having a single definition of a mandatory licensable HMO applicable across all of England or Wales would be very inflexible and would not take account of local situations. Therefore, the 2004 Act empowers individual local housing authorities to seek to license HMOs with other descriptions. This is called additional licensing. A local authority might choose to licence HMOs with four or more occupiers, or even all HMOs. Additional licensing schemes can apply to s.254 HMOs or s.257 HMOs, or to both types. Such schemes can also be limited either geographically, by property description, or both. In order to set up such a scheme, the relevant local authority must hold a consultation process and may need to gain approval from the Secretary of State (SoS) if that is required.

Additional licensing schemes are designated for five years at a maximum or such lesser period as the local authority initially designated. Schemes can be revoked by the local authority prior to that point, but if they are not they will end automatically at the end of the original designation period and the local authority will need to hold another consultation and approve a new scheme. In practice, this will need to be done in advance of the old scheme ending as otherwise there will be a short period between the two schemes in which licensing will not be required. There have been a number of cases where the local authority has failed to match the two points up correctly and so one scheme has ended without a new one beginning. This does not cause any particular problems for landlords, but certainly makes the situation more confusing. It has never been clarified if a licence issued under a designation which extends, or could extend, beyond the end of that designation would continue to be effective under a subsequent designation, either one which is contiguous and commences as the first one ends or one which starts some period of time afterwards.

Selective Licensing

Selective licensing schemes allow local authorities to introduce licensing of all landlords in a designated area, even where the properties are not HMOs. As with additional licensing these schemes can be limited to a specific area or by a property description or both. Again, in order to set up such a scheme, the local authority must hold a consultation process and gain approval from the SoS if required.

Selective licensing schemes also last for five years at a maximum or such period as the local authority initially designated if it is less. They can be revoked by the local authority prior to that point, but if they are not they will end automatically and the local authority will need to hold another consultation and approve a new scheme. In practice, this will need to be done well in advance of the old scheme ending as otherwise there will be a short period between the two schemes in which licensing will not be required.

In April 2010, changes were made to the process for granting additional licensing in England. Previously, the process required a consultation and proposal to be placed before the SoS for him or her to accept or reject. While the consultation and consideration must still occur at the local level, it is no longer necessary to get approval from the SoS for additional HMO licensing schemes. The need for approval had already been waivered in Wales. This has, unsurprisingly, led some local authorities to forge ahead with substantial licensing schemes.

Setting Up an Additional HMO or Selective Licensing Scheme

Any local authority that wants to have a licensing scheme beyond the mandatory scheme is required to have a consultation. There is also the need to obtain approval from the SoS or, in Wales, the relevant Welsh Minister. However, it is possible for that approval to be given in advance on a blanket basis for designations that fall within a specified description and this has occurred in both England and Wales.

In Wales, any local authority can make a selective licensing or additional HMO licensing designation of any size within their area of responsibility without further approval from the Welsh Minister.[7] In England, a similarly relaxed position pertained for some time and local authorities could make designations for additional HMO or selective licensing without further approval.[8]

The rules on approvals were tightened for England in April 2015.[9] The new approval scheme allows that any additional HMO licensing scheme will be approved as long as it has been the subject of a consultation of not less than ten weeks in length. It is now the case that Councils cannot create unrestricted selective licensing schemes without the consent of the SoS. They can create a scheme which affects less than 20% of their geographical area and less than 20% of all private sector rented homes in their area. Anything more than this will require consent. Additional licensing schemes remain unrestricted. The process of approval is inevitably somewhat politicised. However, there have been approvals across party lines. It seems that in England there has been a move from central government to look sceptically at schemes which cover all of a local authority's area of operations. In practice, this may be right as it seems unlikely that a local authority is able to provide evidence across the entire sweep of its area of operations that would meet the criteria for a licensing scheme. Frustratingly, from a local authority perspective, the SoS appears not to give reasons for decisions to reject or approve schemes and appears not to even record them in writing. Traditionally, there is no clear duty in law for decision-makers to record or provide reasons for making a decision but the Courts have leaned progressively towards the view that there should be and they have certainly held that reasons can be required where they are necessary to allow scrutiny by the courts of a decision.[10] Given that a decision to refuse or approve a scheme lies entirely within the purview of the SoS, but presumably has to be based on the statutory criteria and a review of the local authority evidence, then it would seem that a refusal of consent, which must necessarily involve a different view of the evidence and criteria than that taken by the local authority, should require the SoS to record the reasons for that decision.

In Wales, selective licensing schemes are not subject to the limits associated with England. However, in practice, they have been made far less attractive by the advent of mass landlord licensing in Wales. At the time of writing, no Welsh local authority operates a selective licensing scheme.

Nature of Consultations

The need to consult before making a licensing designation has been a significant problem and many local authorities have not done a great job. The government has produced guidance on what a consultation should look like and the courts have also given important opinions in this area.[11]

The consultation must be a real consultation which is fair and genuinely intended to solicit feedback on the planned designation with the aim of validating or improving it. This means that the consultation should follow the, so-called, Sedley Criteria.[12] These are that:

1   The consultation must be carried out at a time when the proposed designation is still at a formative stage;

2 The local authority must give sufficient reasons for the making of the designation to allow for intelligent consideration and response;
3 Sufficient time must be given for consideration and response; and
4 The responses to the consultation must be conscientiously taken into account in finalising the designation.

The second of the Sedley Criteria has been amplified by the courts in critiquing a consultation which consisted of a "listening exercise" that did not contain any substantive proposals.[13] The third criteria has also been dealt with in England by the approvals Order made by the SoS which stipulates that approval for a licensing scheme is only automatically available where there has been a minimum ten-week consultation period. Therefore, most consultations now follow this timeline. That ten-week period can be made up of more than one consultation but it cannot include any period of consultation in which the other Sedley Criteria are not being satisfied.[14]

However, a consultation is not a local referendum. There is no requirement that a majority of responses, or necessarily any of them, should be supportive of a designation. That said, if every response was to be against a designation, then it would be a brave local authority indeed that proceeded to make a designation without at least a careful consideration of the objections and whether those objections contained some reasonable point that they should take into account. Simply ignoring responses and not considering alternatives to a proposed licensing scheme is not permitted within the legislation[15] and is also contrary to the principles of a properly conducted lawful consultation.[16]

The area embraced by the consultation is also important. The local authority is required to "take reasonable steps to consult persons who are likely to be affected by the designation".[17] The High Court has held that this may well extend beyond the immediate area that the licensing scheme is intended to cover and may even extend beyond the area of responsibility of the local authority and into areas controlled by other local authorities.[18] This is because a designation, especially one that is covering a substantial part of the local authority area, is likely to affect those outside that area. Either because they operate businesses dealing with properties which are affected by the designation or because the designation will have knock-on effects for their area. The exact scope of the consultation and how widely it needs to extend beyond the planned designation area will depend on the local economic conditions and the size of the designation area. So, a larger designation area is likely to need a wider consultation area as is a designation in a regional centre which might have far-reaching effects in the surrounding area. There is substantial guidance from the government on good decision making which is directly applicable to consultations and designations and any local authority officer considering how to conduct a consultation and make a designation would be well advised to consider it carefully.[19]

However, many landlords think that they can judicially review a licensing designation on the basis that they do not like it. In practice, it must be remembered that substantial elements of a decision to make a designation fall within the discretion of the local authority. A judicial review is available if it can be shown that the consultation was not conducted properly or lawfully or if the decision to make a designation is irrational. The fact that a different council might have made a different decision or that the decision is not a good one for a group of landlords or agents is not sufficient. Often the threshold for having the decision replaced by the court is not met. It should also be borne in mind that if a local authority has their consultation impugned, the solution for that local authority is usually to simply hold a further consultation. So, in practice, a judicial review against a licensing designation often achieves little more than a delay in it being made. This may be useful in specific cases to allow for landlords to reconfigure their businesses or to allow for more time to offer alternatives to the local authority but there is no certainty that legal action will lead to a designation disappearing altogether.

A designation can be of any length of time specified by the local authority when it is made, up to a maximum period of five years.[20] There does not immediately seem to be a power to change a designation once it has been made to cover a different area, different property type, or a different time period.

Once a scheme comes to an end, there must be a full consultation on any new scheme. This means that there has to be substantial forethought and careful timing over the new consultation and designation process if the aim is to have the old and new designation mesh seamlessly together. In practice, not every local authority has managed this and there have been a number of designations that have lapsed with a new one coming into effect a few days or weeks later leaving a gap during which there is no requirement to licence some properties.

Notification of a Licensing Designation

The 2004 Act sets out that a licensing designation must be notified in a specified way.[21] More detail of the exact form of a notification is to be found in supplemental regulations.[22] These notification requirements come into force as soon as a local authority makes a designation or, where it needs to be confirmed by the SoS or Welsh Minister, when that confirmation occurs. In some cases, local authorities have not been properly prepared for this step and there is a delay between the formal making of the designation through the local authority decision-making process and its notification.

A formal notification must state:

- that the designation has been made;
- whether or not the designation was required to be confirmed by the SoS or Welsh Minister and that it has been confirmed;
- if confirmation was not needed, then the notification must state that it was made using an existing general approval already made by the SoS or Welsh Minister and must further give details of that general approval; and
- the date on which the designation is to come into force.

It is open to government to specify further information which needs to be given in a designation notice by way of regulations and these require that a notification also includes:

- a short description of the area covered by the designation (often this will refer to a map as well);
- the name, address, telephone number, and e-mail address of:
  - the local housing authority that made the designation;
  - the premises where the designation may be inspected; and
  - the premises where applications for licences and general advice may be obtained;

    In practice, these contact details tend to be the same for all three elements;
- advice to any landlord, person managing, or tenant within the relevant area to obtain advice from the local authority as to whether they are covered by the designation; and
- warnings, including details of the criminal prosecutions that might be taken and their consequences, if a licence is not obtained for a property affected by the designation. It is still frequently the case that local authorities do not do this correctly and fail to specify the penalties of prosecution properly.

The regulations are also very specific about the manner in which the notification must be published. The timescales are tight in that the notification process must commence within seven days from the decision to make the designation. This is often a problem as in practice the notification process needs to be pre-prepared before the decision is made and so it is necessary for officers to assume that the decision-makers (usually elected representatives) are going to agree to make the designation.

Once a designation is made, then within seven days of that date a local authority must do all of:

- place the notice on a public notice board at one or more municipal buildings within the area covered by the designation or at the closest municipal building outside the relevant area if there are none within the area covered by the designation;
- upload the notice to the local authority's website; and
- arrange the publication of the notice in the next edition of at least two local newspapers circulating in or around the designated area.

While the notice in the two local newspapers does not need to be published within seven days as it merely has to be arranged it does need to make the next edition and so it may be necessary to have this arranged prior to the decision being made. The notification needs to be published in a further five editions of the same two newspapers at an interval of not less than two weeks and not more than three weeks.

In practice, these notification requirements are becoming increasingly hard for local authorities to meet and have not properly kept up with the changes in local publishing caused by the internet. A number of areas simply do not have two local newspapers covering them anymore and local newspapers are also moving to online (sometimes online-only) publication and have frequently ceased to have "editions" in the classical sense. It can therefore be difficult to find two suitable local papers to place a designation notice in. The regulations also do not take proper account of the increased use of social media and other communication methods. They are therefore sorely in need of updating to focus more clearly on online publication and notification.

As well as publication in the press and on noticeboards within seven days, the local authority must also within two weeks send a copy of the designation notice to:

- anyone who responded to the consultation about the proposed designation;
- organisations which represents the interests of landlords, tenants, estate, letting or managing within the area covered by the designation; and

- organisations which provide advice on landlord and tenant matters such as law centres, citizens' advice bureaux and the like.

The issue of notifications has come up to some extent in the courts in a case in which a district council sought to prosecute for a failure to have a selective licence.[23] The landlord argued that he had a reasonable excuse for not having a licence as he did not live in the area and the local authority had not sufficiently publicised the scheme to him (as he was not aware of it) and so they had not taken all reasonable steps to procure that applications for licences were made to them as they are required to do.[24] While the magistrates were initially persuaded by this argument, the Divisional Court rejected the argument on appeal. The council had fulfilled their obligations to publicise the designation for selective licensing and had complied with the legal requirements. If it had been the case that they had not fulfilled their notification obligations, then there might have been a reasonable excuse defence available. However, there was no evidence of such a failure and it was not possible to proceed backwards from the fact that the landlord was unaware of the designation to then conclude that the council had got things wrong. Therefore, a local authority is not required to do any more than fulfil its statutory publication requirements whatever they might be. But if it fails to do so, then there is a real risk that it will find it impossible to prosecute landlords who can reasonably claim to be unaware of the designation having been made.

Review and Revocation of Designations of Licensing

There is a general requirement for local authorities to review a licensing designation made by them from time to time. In practice, this obligation is not commonly fulfilled in any substantive manner and there is no statutory requirement to carry out such a review at any particular point.[25] On a review, the local authority is empowered to revoke a licensing designation if they consider it appropriate to do so.[26]

When a designation is revoked that revocation takes effect from the date specified by the local authority when it makes the revocation decision.[27] Just as with the making of a designation, there is a requirement to provide a notification of the revocation of a designation[28] and the specifics of this are set out in regulations.[29]

A revocation notification must contain:

- A brief description of the area covered by the designation which is being revoked;

- A summary of the reasons for the revocation;
- The date from which the revocation will take effect; and
- The name, address, telephone number, and e-mail address
  - Of the local housing authority that revoked the designation; and
  - Where the revocation may be inspected.

This notification must be published in a similar way to the notice of making the designation in the first place, although the requirements are a little less onerous. For a revocation, the notification must be published within seven days of being made in all of the following ways:

- By placing the notification on a public notice board at one or more municipal buildings within the area covered by the revoked designation or at the closest such buildings outside the area if there are none within the designation area;
- By publishing the notification on the authority's website; and
- By arranging for notification to be published in at least two local newspapers circulating in or around the area covered by the revoked designated in the next edition of those newspapers.

As with the publication process for making a designation, this process is somewhat outdated and could be changed by updating the relevant regulations.

Revocation seems at first sight to be an all or nothing arrangement. The scheme as a whole either continues in place or it is fully revoked. However, the partial revocation has been used by Westminster City Council when a part of its additional licensing scheme was revoked only two months after its introduction to remove parts of the scheme relating to s.254 HMOs while retaining parts relating to s.257 HMOs. It is unclear whether this partial revocation was actually lawful. A similar issue arose in Tower Hamlets Council where the Council amended its existing selective licensing designation after it had been made, which it described as a "small administrative change".[30]

The Housing Act 2004 only allows for revocation of a scheme following a review rather than variation or partial revocation. Whether the power to revoke also embraces a power to partially revoke or vary a scheme is unclear and ultimately the Courts and Tribunals will need to determine this point. The authors doubt whether this can in fact be done and it is possible that schemes which have been partially revoked are no longer lawful. The more interesting question is whether the courts would find that the revocations and alterations have therefore never taken place and so the schemes are in effect as they were originally specified or, worse, whether the attempted revocation or alteration leads to the entire scheme being void meaning that those schemes are now entirely defunct.

Applicability of Licensing and the Duty to Licence

It is frequently the case that the applicability of licensing is misunderstood. Once a licensing scheme comes into force in an area, then any property in that area which falls within the definition of the scheme must immediately be licensed. To that extent, licensing is retroactive in that it applies to already existing tenancies and landlords are effectively under and obligation to be aware at all times of licensing schemes that might affect them, not merely to check on the renewal or commencement of a tenancy.

By the same token, however, a local authority is under a duty to make suitable arrangements to effectively implement a licensing scheme in their area and also to ensure that licence applications are dealt with inside a reasonable time.[31] It is often the case that a local authority will implement a licensing scheme at a time when they are not fully prepared to actually process the applications and it is often the case that online application systems are non-functional or application forms unavailable until very late in the day. They cannot therefore possibly be in a position to comply with their statutory duty to properly implement a licensing scheme. The local authority will then suggest that landlords will be given a 'grace period' during which they are able to seek a licence but where the local authority will not pursue prosecutions. This is a very high-risk approach for a local authority to adopt. Local authorities, like all prosecutors, have a discretion whether or not to commence a prosecution in a particular case. However, that discretion is supposed to be exercised on a case-by-case basis and decision-makers should not adopt rigid policies which prevent them exercising their discretion or excessively limit it.[32] In other words, a decision-maker should make a decision in each case that comes before it by exercising its judgement and not replace judgement with a written policy. In particular, an enforcement body owes a duty to the public to enforce the law and is not permitted to simply adopt a general policy that states that it will not do so.[33] A local authority may adopt a lenient policy towards those who have not obtained a licence for a period of time but it would be wrong to simply state that it will not prosecute at all. As well as being potentially unlawful, the suggestion of a 'grace period' also leaves landlords in England in a very risky position because of the various direct enforcement powers that tenants have in respect of landlords who have not applied for a licence. Tenants will not be bound by any leniency or grace period offered by a local authority.

The Application Form and the Application Fee

It is up to each local authority to decide what a licence application form will look like.[34] However, that does not mean that they have total freedom as to what information or declarations they require in an application. There are certain things that have to be in a licence application and these are set out in regulations.[35] These include:

- Contact details for the applicant, proposed licence holder, and persons managing or having control of the property;
- The address of the property the licence is being applied for;
- The approximate age of the original construction of the property;
- The nature and structure of the property;
- The current occupation of the property;
- Details of any other properties which the proposed licence holder holds licences for; and
- Declarations as to fire safety, furniture, and gas safety.

In addition, all applications need to include details of relevant convictions and findings of unlawful discrimination as well as any licence refusals or revocations made against the proposed licence holder or the proposed manager.

The scope of the local authority to go beyond this list is pretty limited. The High Court has held that the list of items in the regulations is the maximum extent of what the local authority can require landlords to provide.[36] These regulations were amended in England to further restrict the information required for a renewal to an even shorter list.[37] In England, on a renewal, the local authority is only entitled to seek details of the applicant, proposed licence holder, and manager and the address of the property itself. The High Court has therefore said that a local authority was wrong to ask for details of the occupants on a renewal as this requirement had been specifically removed in order to reduce unnecessary burdens on landlords.

As well as the specific list of items set out for the application form any applicant is obligated to write to their mortgagee, any superior landlord, and any person with an interest in the property and inform them that the application has been made (Figure 3.1).

Licensing of Properties 41

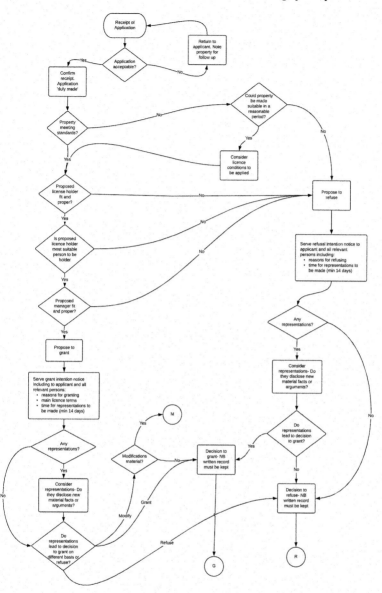

Figure 3.1 Licence application flow.

Duly Made

A landlord is compliant with the legislative requirements on licensing for all purposes as long as an application for a licence has been "duly made", even if the local authority has yet to consider the application.[38] The key question in most cases is whether or not the application form submitted to the local authority has been properly completed and has all the appropriate information with it. If the local authority has sought information that it is not entitled to, then a failure to provide that information cannot then lead to the local authority suggesting that the application is not duly made.[39] In relation to other similar matters, the courts have taken the view that an application must meet the requirements of the relevant regulation which sets out what must be provided in order to be "duly made". So, an application which is missing one or more of the items required by the regulations will not be duly made. However, a local authority should allow a reasonable time for an application to be put in order if it is defective before treating an application as not being made.[40] In a practical sense, local authorities should respond to applications that are missing parts identifying what is missing and giving a reasonable time to correct the defects. After that time period has elapsed, they are entitled to treat the application as not having been made and act accordingly. This area is of particular uncertainty where local authority licence application websites are defective and, for example, fail to process applications or do not process the payment properly. In these cases, is the application duly made? Even if a local authority were minded to overlook the situation and give a landlord the benefit of the doubt, a tenant might be less forgiving and may, at least in England, seek to recover their rent by way of a Rent Repayment Order (RRO).[41]

Data Protection and Licence Applications

With the advent of the General Data Protection Regulations (GDPR),[42] there has been increasing interest in how data is collected and stored and the justification for doing this. In general terms, a local authority is entitled to process the information it is seeking in a licence application form on the bases that it is under a legal duty to do so and that it is necessary for a task carried out in the public interest or on the basis of official authority vested in the local authority.[43] However, these duties do not remove the more general obligation on the local authority as a processor of personal data to comply with the data protection principles. Therefore, they must limit personal data collection to that which they need and also keep it up to date and reasonably secure. However, not all the data that is being collected is necessarily personal data. It would be hard to claim that the address of a landlord's rental property was personal data, for example and it is something that is essentially available on the Land Registry to any member of the public. It is also open to dispute to what extent any of a landlord's data is actually personal data as it is being used in connection with the operation of a rental property which is a business. However, it has generally come to be accepted that personal information such as home address and personal contact details remain personal information, even if used within a business context. So, while not all data is personal data, there is also worth noting that the right to process data is based on a statutory authority. If a local authority is collecting more data than the legislation itself allows, then that data is likely to fall foul of data protection obligations if it is personal data. In general, local authorities should be cautious to limit publication of personal data of landlords such as their home addresses and other details, doing only what the legislation requires, but they can be more free with the details of landlords incorporated as businesses.

How Much?

The fee charged for a licence causes irritation to many landlords. The legislation specifies that a local authority may set a fee which covers all of their costs in carrying out any of their functions relating to HMO or selective licensing. In principle, this means that all the costs of educating landlords and tenants about a licensing scheme, all the costs detecting unlicensed landlords, and all the costs of enforcement activity can be charged as part of the licence fee.[44] As against that this is the only charge that a local authority can make and they are not permitted to charge for anything else such as copies of licences or applications to vary the terms of the licence. This point has been endorsed by the First-tier Tribunal (FTT).[45] However, the High Court has made clear[46] that local authorities must also take into account the EU Services Directive.[47] The scope of this directive in relation to fees was clarified by the European Court of Justice (ECJ).[48] The ECJ was ruling on the licensing of sex shops but held that charging a fee for authorisation under a scheme regulating sex shops could only require payment of an application fee which covered the costs of processing the application itself. An additional second stage fee charged to successful applicants relating to the costs of administering the scheme as a whole was legitimate, but the authority could not demand the entire fee upfront. The

44   Licensing of Properties

High Court applied this same principle to HMO and, by implication, selective licensing. This means that, notwithstanding the wording of the legislation, a local authority cannot charge a licence application fee that exceeds the cost of processing the licence application itself. It may be possible for a local authority to charge a full fee and then refund unsuccessful applicants the component that does not relate to the application process or to charge a two-stage fee but the court did not make any ruling on whether either of these is likely to be acceptable. However, a two-stage fee does not appear to meet the requirement of the legislation for a single application fee so it is not certain that this second option is acceptable at all.

The form of an application is critical as once the application has been made to the local authority the obligation to licence has been fulfilled until such time as the licence application is determined and so there is a complete defence to any attempted prosecution by a local for any period during which a licence application is before them. However, this defence only applies if the licence application has been "duly made".[49] This means the application must be properly made in the manner reasonably specified by the local authority and the proper fee paid.

How Quickly?

Some local authorities can be very slow in dealing with licence applications. All local authorities are under a duty to ensure that applications for licences are dealt with in a reasonable time.[50] There is no clear legal position in England and Wales on what a reasonable period is to determine a licence application, but some local authorities have taken 24 months or more on occasion and this is unlikely to be acceptable. In Northern Ireland, the legislation requires that an HMO licence application is determined within three months and, if this does not happen, the licence is granted by default.[51]

However, there is little value in a landlord pursuing legal action to force a decision. Taking a case in judicial review to require a local authority to make a decision is not likely to provide a great deal of benefit. Perversely, it is worth noting that delay is arguably in the interests of the landlord as long as a licence has been applied for. Once a licence is applied for, the landlord is safe from prosecution and other penalties but is not bound by any of the conditions and limits relating to a licence, as it has yet to be granted. Therefore, during this *interregnum*, the landlord is essentially free to use the property as they wish without troubling themselves by the eventual limits that will be imposed on the property when the local authority finally grants a licence. Of course, they may find themselves in difficulty if they have used the property carelessly and are then trapped with conditions on the licence that they are unable to meet because they have allowed over-occupation or occupation far outside the terms of any conceivable licence. However, a landlord who had made a reasonable guess at the eventual terms of a licence and then used the property in accordance with that guess would be unlikely to face a penalty if the terms of the final licence did not accord with the use they were making and would have a good case in asking for the licence to be varied on a short-term basis to allow them to bring their property into compliance with the final conditions applied.

While there is no specific provision in the 2004 Act as to how quickly a licence must be decided, it does potentially arise elsewhere. The EU Services Directive has been enshrined in UK law[52] by way of regulations and has been extended beyond the departure of the UK from the European Union (EU).[53] The relevant regulations requires that:

- Applications for a licence or other authorisation are processed as quickly as possible and within a reasonable period;
- That the processing period is fixed and made available to the public; and
- That the processing period may be extended once for a limited period, but that extension must be notified to the applicant in writing in advance.[54]

Moreover, the Directive states that where a local authority fails to process a licence within the published time period or a notified extension, then the approval or licence is deemed to be granted by default unless there is an overriding public interest in automatic approval not being granted. This would seem to suggest that a number of local authorities which are not granting licences promptly should in fact be deemed to have granted a licence.

As a counter to this, it is common for local authorities to assert on their licensing website pages that they do not allow for deemed automatic grant of licences on the basis that there is some form of overriding public interest. The reason given is that it is necessary to protect public safety to have an inspection of the property prior to granting a licence. However, this is a difficult position to arrive at. Firstly, it is hard for a local authority that has delayed in granting a licence for many months to assert that it has been concerned for public safety during that period. Given that a landlord who has applied for a licence is subject to no substantial control during the period of consideration, then a local authority is in a weak position to assert that they are protecting the public by waiting until they inspect the property. Second, not all local authorities inspect properties before granting licences. A local authority that does not inspect at all is in an even weaker position than one that inspects on a delayed basis as the public interest in having an inspection before licence grant cannot apply. Finally, it might in fact be more protective of the public for a local authority to grant licences based on information provided in the application and based on a set of conditions that are standardised and then inspect properties and prosecute for failure to adhere to the relevant conditions or revoke licences where the property was not at the standard suggested by the landlord's application form.

While the above argument has not been endorsed by a court or tribunal, it is certainly questionable whether local authorities can justify extended delays in licensing approvals. The continued extension of these scheme and, in England, the rise of RROs means it is only a matter of time before the issue comes to a head.

Extended delay is also somewhat unfair on tenants. While there is a register of licensed properties, there is no register of applications that have been submitted and are awaiting processing. In Northern Ireland, for example, HMO licence applications are required to be published online,[55] although this is a much smaller jurisdiction and there is a statutory requirement for public consultation on these applications. In England and Wales, with no such register available, a tenant cannot be sure that the property is unlicensed as the licence application may be pending before the local authority. Given that a tenant living in an unlicensed property, that does not have a pending application, cannot be served with an s.21 notice and, in England at least, can seek a RRO, being aware of the progress of a licence application matters.

Understanding Standards and the Licence Grant Process

When deciding whether to grant a licence, the authority must consider whether the property and its management are suitable in the case of an HMO licence[56] or, for selective licensing, whether the management is suitable.[57] The condition of the property and its suitability are not relevant for selective licensing applications, only its management.

HMO Property Standards

When considering whether a property is suitable for an HMO licence, the key test is whether the licensed property is "reasonably suitable for occupation by not more than the maximum number of households or persons" set out in the application. Alternatively, the authority must consider whether the property is suitable for some other number of persons, whether greater or lesser than the number sought in the licence. Therefore, it is entirely possible for an HMO licence to be granted for more people than was originally requested in the application. Equally, the local authority may conclude that the property is not suitable for the number of people sought but is suitable for some lesser number and may propose to grant a licence for that number instead.[58]

When considering whether a property is suitable for a specific number of people, the local authority is not allowed to conclude it is suitable if it does not meet nationally prescribed standards. But just because a property does meet national standards that does not automatically make the property suitable. So, national standards should be seen as a minimum quality level below which a property must never fall but simply meeting those standards is not sufficient to make it suitable.

The national standards have been prescribed in regulations.[59] Initially, the regulations for Wales and England were slightly different. However, the regulations for England were amended to clear up an element of impracticability around the placement of sinks in rooms and ended up mirroring the Welsh regulations.[60] Both sets of regulations were also further amended to set standards for s.257 HMOs.[61] There have been considerable challenges in using the standard regulations in practice. This is because they are both highly prescriptive, in the sense that they have a series of very detailed standards for things such as the number of toilets and bathrooms while simultaneously being very loose about other matters such as kitchen size.

Local Standards

Some local authorities have sought to fill in the gap in the national standards by seeking to publish their own standards for HMOs and to treat them as a series of fixed standards which must be met. If a property does not meet what are often quite prescriptive standards, then the licence is refused. The Upper Tribunal (UT) has held that this approach is entirely incorrect.[62] Councils cannot set fixed standards, that is for the government to do and no power has been created to allow local authorities to do this, even where they consider that government standards are unclear or lacking in some respect. So, local authorities can have detailed guidance and can apply it in a relatively rigid manner but they must be reasonably flexible in their application and not use standards as a means of avoiding a holistic consideration of whether the property is suitable for the use sought.

Management and Suitability

The second element of the consideration for an HMO licence and the only element for selective licences is whether the management of the property is suitable and whether the persons carrying it out are fit to do so. Fitness is discussed below but the issue of management arrangements is one that many local authorities do not pay sufficient attention to. There tends to be an assumption that any management arrangement is acceptable without looking at the details of that arrangement and how it operates. Unless the nature of the arrangement and the level of responsibility is considered, then it is hard to be sure that the arrangement is suitable. However, local authorities are in something of a bind in this area. They are only permitted to ask for information that is on the list of items in the regulations. This list includes the identity of the managers but it does not include information about the management arrangements or information about the staffing of the manager if it is a company and how their internal arrangements work. This is an area in which local authorities need to develop their local intelligence arrangements so that they know about common local managers and how they operate.

Fitness and Suitability

One of the key issues the local authority must consider is whether the proposed licence holder is fit and proper to do so.[63] In general, the local authority must grant the licence to the person who is most appropriate to hold it and this person will usually be the person who holds a freehold or long lease interest in the property. The grant of a licence to another person is perfectly possible but it must be justified.

Fitness of the licence holder and manager is an overall consideration based on the history of that person. Offences relating to property, honesty, and sexual offences are all of great importance. The test is also relational so a person can be unfit if their spouse or a direct family member has been convicted of an offence and a company can be unfit if a director has been convicted. The fact that a related conviction exists is not an immediate determinant of unfitness. The test is relational in order to stop unfit persons from appointing a false person of straw to stand in for them and obtain a licence while they continue to run things in the same way as before from the shadows. The fitness test is applied in the similar way to managers and managing agents.

When considering offences, it was generally believed that local authorities were bound by the provisions which allow for rehabilitation of offenders. A person who has only been subject to a fine is considered rehabilitated 12 months after the date that they were convicted, not the date of commission of the offence or the date of sentencing.[64] Once a person is rehabilitated, they must be treated as if they had not committed the offence at all and if asked they are allowed to state that they have not committed the offence.[65] This applies to any hearing before a court or tribunal as well as any local authority making decisions by virtue of legislative provisions.[66] So, it was thought that local authorities could not consider an offence committed by an individual which was punished by a fine more than 12 months after conviction. Likewise, a tribunal is also unable to consider such offences. However, this limit only applies to individuals and so companies do not enjoy the same protections. It is open to the government to disapply the rehabilitation provisions in relation to information being sought for specific purposes but this has not been done in relation to Housing Act 2004 matters. The belief in rehabilitation was so strong that it was even repeated in central government guidance. However, the Court of Appeal has not agreed. It has held that while the existence of a conviction may not be brought into consideration, the facts that gave rise to it may be as long as they are relevant. That may also include facts relating to prosecutions that were not commenced or which were discontinued. The Court has further held that this was not a matter just for courts and Tribunals to consider but that a local authority could make the same consideration. However, that does not mean that all such facts should be considered. Instead, the local authority, or Tribunal as appropriate, should first consider whether it was inclined to admit this information and give the person involved a chance to make their case as to why the information should or should not be considered. Once that decision has been made, then a full consideration on fitness should be made, including or excluding the relevant material, as decided.[67]

Just because a person has been convicted of an offence and regardless of whether they have been rehabilitated does not automatically make them unfit to hold a licence. In practice, a local authority should be looking at a number of factors in relation to people who have been convicted of an offence. These should include:

- Relevance
  - There are a list of offences which are subject to consideration, including housing offences and those involving dishonesty.

However, not every offence that falls within the relevant categories may in fact be relevant to the situation. So, while a single dishonesty offence may not be relevant, if it involved multiple frauds a range of issues that might be relevant. The facts are also relevant and so a single fraud involving housing-related benefits may be of very great relevance to a determination of fitness to manage a property.

- Seriousness of the offence:
  - Was the offence one which was committed caused by a particular situation arising (such as speeding or HMO licensing) or did it require an element of pre-planning?
  - How serious did the court consider the offence to be? How severe was the sentence by comparison with the maximum sentence available for the offence? What were the sentencing remarks (if any) of the judge or magistrates?
- Rehabilitation:
  - How far through the rehabilitation period is the offender or has rehabilitation already occurred? It is a continuum of risk not a position whereby they cease to be a risk at the moment they are rehabilitated or at some other date. An offender who has committed a minor offence some time ago is likely to be fit.
  - Has the offender done anything to demonstrate that they have reformed? For example, taking relevant courses in property management.
- Mitigation:
  - Are there any options to reduce or mitigate the risk? For example, by requiring a professional property manager or imposing a licence condition relating to management that would enable better review of the standard of management.

However, it is not open to a local authority to grant a shorter licence to a landlord merely to allow an existing conviction to become spent.[68] The fact that a conviction is spent is relevant but it is not the determining factor for licence length, or really the determining factor for fitness either. It may be the case that a conviction is relevant to length of licence granted but only if there a clear and direct link between the conviction and the length. That might occur if the breach or offending was ongoing or if there was a high chance of repeated offending. However, merely waiting for a conviction to be spent should not be relevant.

In England, the fitness tests have been further extended.[69] So, it is now also open to a local authority to consider whether a person is fit if they need, but do not have, leave to enter into or remain in the United Kingdom or if they are insolvent or an undischarged bankrupt. These tests must be considered alongside the Provision of Services Regulations 2009 however. These make clear that restrictions or prohibitions on licence holders which are based directly or indirectly on nationality or place of establishment (for companies) are unlawful. They cannot require any person or staff member to be resident in the UK either.[70] Therefore, there is a tension between a fitness consideration based on ability to enter the UK and a set of regulations prohibiting consideration of nationality or residence. Local authorities will need to be sure that they only consider right to enter or reside only and do not at the same time breach the restriction on nationality or residence.

In addition, a further test is added which allows local authorities to consider evidence of a person being served with civil penalty notice due to a breach of the Right to Rent legislation.[71] While this provision purportedly applies in England and Wales, the Right to Rent provisions are only in force in England and so its application in Wales is limited.

Finally, local authorities in England should be considering data held on the Rogue Landlord and Agent database which suggests that the licence holder or manager are a rogue landlord or agent. In particular, any landlord or agent that has been banned for operating, and who will accordingly be listed on the database, is automatically deemed to be unfit.[72] However, it should be remembered that offences might be listed on the database which are spent because the offender has been rehabilitated. The fact that an offence is listed in the database for a period which exceeds the rehabilitation period does not mean that the provisions relating to rehabilitation do not apply and so a local authority may still not consider those offences when deciding whether a person is fit for the purposes of granting a licence albeit that they would be entitled to consider the events which led to those offences being committed.

Management arrangements must also be suitable. This means that all the above fitness considerations should be applied to the agent and their key personnel. But it also means that the entire scope of the management arrangements should be considered. Not all local authorities give full consideration to these provisions. However, they should be considering whether agents who are responsible for managing an HMO have sufficient authority to carry out works that are likely to need doing, for example, or whether they are forced to seek permission from some other person or organisation. In the latter scenario, it is questionable whether the management arrangements would be suitable as the agent would have no capacity to carry out necessary work without reference to someone else and so it would be difficult to hold them liable for any failures.

When considering companies and individuals, there is also a relational aspect to the tests. So, when considering the fitness of an individual, it is also relevant to consider the fitness of those people associated with them and any relevant unspent offences they might have committed. Likewise, it is appropriate to consider employees of a company and relevant unspent convictions that they might have as well. However, as with convictions generally that does not mean that any company which has an employee who has a relevant conviction or any person with a relation who has been convicted is automatically unsuitable. The provisions are directed at preventing those who have been convicted from putting forward their relations or a company as a false front and simply carrying on their business from the shadows using a proxy as their public face. So, for an individual, it will be necessary to consider their relationship with the offender and the level of control that person might exercise over them. For a corporate, it will be necessary to look at the appointment of the individual in the company and what level of control they exercise over the management of property. So, if they are the sole director, then their convictions are likely to be of great concern, whereas a bookkeeper is unlikely to be particularly relevant to a licensing application. These considerations should also be made alongside the various points already discussed as to the seriousness of the original offence, the level of rehabilitation, and any mitigation measures that might be possible.

All of these considerations remain subject to the other crucial limitation, that a local authority can only ask for the information set out in the relevant regulations.[73] So, their ability to explore the above issues is limited unless they have derived information that would justify asking for more details. So, it is not open to a local authority to demand that individual agents, for example, provide checks from the Disclosure and Barring Service, for example as this is not covered by the regulations.

Licence Conditions

It should be remembered that all licences have conditions attached to them. The main one is the number of persons permitted to occupy the property. Overcrowding a property is a serious offence with an unlimited fine. Breach of other licence conditions is an offence which also carries an unlimited fine.

However, it should be noted that local authorities have limitations in relation to conditions.[74] Firstly, a licence cannot have any condition within it which seeks to impose on a person other than the holder of the licence unless that person has consented. That means that conditions cannot impose obligations on agents of the landlord or on the occupants of the property without their express consent. In practice, this means that such conditions almost never exist.

Secondly, licences are not permitted to include conditions which are seeking to alter the terms of the occupation agreement that a person occupies a property under. This is a difficult area and there is an aspect in which a large number of conditions might be said to violate it in that many of them have some effect on the manner in which the property is to be occupied. However, the approach seems to be one that looks at conditions that are seeking to create a direct alteration in the terms of an occupation agreement by specifying that it must contain specific terms or that specific terms are forbidden.

The Welsh Residential Property Tribunal (RPT) has held that Councils ought not to adopt a standard set of licence conditions and take a 'one size fits all' approach.[75] That said, there is no doubt that as many houses in a particular area have similar characteristics, there will be a similarity between licence conditions across properties. In practice, many Councils do in fact operate "standard conditions" or have conditions which have the caveat in them "if appropriate". Generally, a degree of pragmatism must hold sway. Local authorities will inevitably have groups of conditions that are similar across groups of properties. The fact that a condition is generic in nature will not mean that it will be disallowed by the Tribunal if it is relevant to the property and the proper management of it. It is a point worth making when considering the quality of a condition and arguing for a change in its scope or emphasis though as the use of standardised conditions would be indicative of a failure to consider the property properly and how those conditions might apply to it. Local authorities must bear in mind that while they might have standard conditions as a starting point, they should be considering these conditions in each case to decide whether they are appropriate to the property that is being licensed.

The 2004 Act sets out specific groups of conditions that can be imposed.[76] One of these is that licences can include "conditions requiring the taking of reasonable and practicable steps to prevent or reduce anti-social behaviour by persons occupying or visiting the house". Often local authorities have reproduced this text almost verbatim in a licence, probably out of uncertainty as to what conditions they could realistically require. However, this creates a nebulous and essentially unenforceable licence condition which merely requires a landlord to take "reasonable and practicable steps" without being very clear what those might be. However, similar wording also exists in the legislation in Northern Ireland. In Belfast, the council has taken the view that this allows them to impose a condition that a licence holder must furnish the council with "an emergency out of hours contact number to allow the landlord or managing agent to be contacted in circumstances where there is anti-social behaviour occurring at the HMO property and the co-operation of the tenants cannot be secured". The High Court in Northern Ireland has upheld the legality of this condition finding that it does fall within the scope of "reasonable and practicable steps".[77] Therefore, it seems likely that local authorities in England and Wales also have much greater discretion in setting licence conditions than many of them have used to date.

There are other conditions set out in the 2004 Act that must be incorporated in a licence, or which are assumed to be there whether or not these are in fact specifically incorporated. These mainly relate to safety requirements such as requiring that gas safety certificates are obtained and to keep the furniture and electrical systems in a safe condition. They also empower the local authority to demand provision of certifications for these items where there is a statutory requirement for a landlord to obtain one.

Local authorities may also use licensing conditions which control the nature of the occupation. Local authorities have been reluctant to do so, often asserting that policing such conditions is impossible. However, the Supreme Court has upheld a licence condition imposed by the FTT which restricted the use of a property to students. This was a somewhat unusual restriction and decision by the FTT in that they implicitly held that students did not require as much space as other tenants on the basis that they would be out of the property more. This characterisation of students may say more about the view of Supreme Court judges than the reality of student life but the point has been made.[78] In practice, the local authority fought hard against this outcome and so it seems that local authorities are reluctant to grant licences which allow a use of a property they would not be prepared to licence normally but then make them suitable by creating a licence condition limiting occupancy to a specific type of tenant.

Conditions in Selective Licences

In relation to selective licences, there is a tendency to assume that all the same conditions as are applied to an HMO licence can be applied in a selective licence. This is not the case. The Court of Appeal has made clear that selective licence conditions should relate to the management of the property and not its condition.[79] This is because the part of the Housing Act 2004 which deals with conditions in HMO licences[80] is worded differently from the part of the Act which deals with conditions in selective licences.[81] In both sections, a licence can have conditions "appropriate for regulating the management, use or occupation of the house concerned" but HMO licences are additionally able to contain conditions allowing for regulating the property's "condition and contents".[82] The Court of Appeal drew a clear distinction between these two different permitted categories of conditions and held that the omission of the phrase "condition and contents" by Parliament was intentional and must denote a specific difference between licensing of HMOs and licensing of other properties. It was also noted that the government is able to make regulations in order to prescribe specific standards as to in selectively licensed property and this would allow local authorities to make conditions which relate to those standards. However, no such regulations have been made. This decision presents a bit of an inconsistency in the minds of some local authority officers. This is because the conditions that must be met for a selective licensing scheme to be granted can be added to by statutory instrument.[83] In England, they were added to[84] in order to state that where an area has a large number of properties in the Private Rental Sector (PRS) and the local authority has reviewed the standard of PRS properties in the area and is proposing to carry out inspections under the Housing Health and Safety Rating System (HHSRS) of a substantial number of those PRS properties then selective licensing is permissible if it assists that process. However, this does not mean that selective licensing can be used to provide for the end product of that process by making provision for property condition in a licence. Standards must be dealt with by using the HHSRS.

Room Sizes

Originally, there was no limits on the size of rooms, they merely had to be reasonable and appropriate taking into account the property as a whole. However, local authorities sought to create their own minimal standards for rooms in an, arguably, misguided effort to fill the gap. This led to cases where the Tribunals found properties to be suitable where one or more rooms were considered by the relevant local authority to be too small.[85] After complaints by local authorities, the government has acted to set out a minimum set of room size standards for England.[86] This operates by amending the 2004 Act in England to insert a compulsory condition into all licences issued or renewed after 1 October 2018. This new condition specifies that rooms used for sleeping can only be occupied by specified number of people if the floor area is of a certain size. For a person over ten years, then the room size must be at least 6.51 sqm for one person or 10.22 sqm for two people. If a room is occupied by a child under ten, then it can be 4.64 sqm, but any room smaller than this cannot be used as sleeping accommodation at all. In addition, rooms of less than 4.64 sqm are required to be notified to the local authority. This only applies to rooms being lived in as sleeping accommodation and not to rooms occupied on a temporary basis by visitors.

In measuring the room size, it is about the wall-to-wall floor area and no regard is given as to whether that floor area is immediately useable. However, any floor area in which the floor to ceiling height is less than 1.5 m is not to be counted when computing the room size. In other words, the average person needs to be able to stand up!

There are criteria permitting a local authority to give a grace period of up to 18 months to a landlord who has breached the room size criteria due to an event that he was not aware of. For example, this applies where a tenant has added a second occupant to a room without consent and will allow a local authority to give a landlord a grace period of up to 18 months to take action to remove the tenant and regularise the situation.

Room size remains a hugely contentious area. The legislation gives only minimal assistance in this area. The law merely states that the property must be reasonably suitable for use by a specified number of persons. Local authority officers frequently lead themselves into error by failing to look at properties holistically and becoming overly focused on single rooms and their perceived deficiencies.

HMO Licence Conditions and the HHSRS

There is an inherent tension between HMO licensing and licence conditions and the HHSRS. Both of them to some degree speak to property standards, albeit in slightly different circumstances. The HHSRS Enforcement Guidance has substantial comment on this issue.[87] However, much of the guidance is unhelpful in deciding whether to deal with issues using the HHSRS or HMO licensing. The only useful commentary repeats the legislation, stating:

Although it is possible to attach conditions to a licence requiring such works to be carried out, section 67(4) provides that authorities should proceed on the basis that generally they should exercise Part 1 functions to identify, remove or reduce category 1 or 2 hazards in the house in preference to imposing licence conditions.

The general tenor of the legislation seems to be that licence conditions should be used more to deal with issues of condition specific to the use of a property as an HMO while the HHSRS should deal with more general issues, albeit with a recognition that risks are different in an HMO.

There is, at least at first glance, an inherent tension between the HHSRS which is a risk-based assessment of the positives and negatives of a property and HMO licensing which is a simpler yes or no answer as to a property's reasonable suitability. This view has been seen in some local authorities who have asserted that properties have rooms that are too small and sought to prohibit their use under the HHSRS and then, having been overruled by the Tribunal, have tried to achieve the same effect by refusing to issue licences on the basis that the same rooms are too small or vice versa.[88] However, that is almost certainly an inaccurate distinction. The assumption that HMO licensing is not risk-based is likely to be wrong. Reasonable suitability implies a consideration of the risks of licensing a specific HMO, especially when the licensing regime is considered in the context of the legislation as a whole. Therefore, the idea that a property has no substantial risks under the HHSRS but is not reasonably suitable for occupation is illogical and, ultimately, not likely to be one that the Tribunals will find themselves agreeing with over the longer term.

The Grant and Refusal Process

Provided that these criteria are fulfilled satisfactorily, then a licence should be granted either unconditionally or with certain specified conditions. These conditions could also include a limit on the number of persons to less than the applicant has asked for. The authority should give notification of their intention to grant or refuse a licence to all relevant parties along with a draft licence and invite representations on those proposals. The local authority must allow a minimum of 28 days to respond but can allow more. Based on the representations, they should then make their final decision.[89] Where the local authority proposes to modify the draft licence based on representations made, then they must repeat the cycle of issuing a draft and allowing representations for a minimum of 28 days.[90] As the draft goes to all relevant parties, this means that it is not just the applicant that can respond. Any person with an interest in the property (which could include mortgagees or superior landlords) should also get a copy of the draft licence and can make representations.

Once a decision has been made, the local authority must serve a notice of intention to grant or refuse a licence on the applicant as well as other appropriate persons. If the decision is to grant, they must also provide a draft licence. There is then a minimum of 28 days during which representations can be made. This can include representations on licence conditions. The local authority must consider these then either grant or refuse a licence or issue a further notice of intention to grant or refuse if the representations have caused them to change their mind. After the final decision is made, there is a short window of up to 28 days during which an appeal against the decision or any aspect of the licence conditions can be made.[91] It is worth noting that an appeal is not just against the decision to refuse a licence but can also be made against a decision to grant one. An appeal against a licence grant can also be made against any of the conditions in the grant of the licence. So, a licence holder might be granted a licence and then appeal the grant of that licence on the basis that they do not accept one or more of the conditions within the licence and want it to be varied.

An appeal against the decision of the local authority is made to the FTT or the Welsh RPT as appropriate.[92] They will consider the entire matter afresh and can do so based on information that was not available to the local authority at the time they made their decision.[93] The tribunal can reverse, confirm, or vary the decision of the local authority and can also direct the local authority to grant a licence on such terms as the tribunal considers appropriate.[94] This does create a small risk for those appealing against a licence condition in that it is possible to be granted a licence by a local authority, to then appeal to the tribunal against that grant on the basis that one or more of the conditions are unacceptable, and then to have the tribunal decide that the licence should in fact be refused altogether or granted with even more onerous conditions.

Liability for HMOs and Licensable Property and the Alternatives

The issue of liability for failure to have a licence, a breach of the HMO Management Regulations, or a breach of a licence condition is complex. The 2004 Act defines the liability strictly and it falls specifically on the "person managing" and the "person having control". Section 263 provides these definitions. It is, however, a little counter-intuitive.

The definition of the person managing is a person who is:

- an owner or lessee of the relevant property; and
- in receipt of the rent from the occupiers either directly or via an agent or trustee;

They are still the person managing even if they are not receiving the rent because they have made an agreement with someone else for that other person to receive the rent instead. This is an anti-avoidance provision to prevent deliberate "round-about" arrangements which throw liability on to shell companies. There is also a further provision that a person managing includes any agent or trustee who is receiving the rent for the relevant premises for the owner or lessee.

Therefore, "person managing" is a very comprehensive definition that covers anybody who is collecting or receiving the rent or could potentially do so, whether they receive it on their own account or for someone else. It will certainly cover a letting agent who collects rent as well as a landlord and will also cover intermediate and superior landlords. So, a person who rents their property to an intermediary who then sub-lets it will still be a person managing. This means there may well be more than one "person managing" the property in some cases. "Person managing" is also somewhat counter-intuitive in that it actually has nothing whatsoever to do with the management of the property but is entirely linked to rent. It is perfectly possible for someone to manage a property in the sense of organise repairs and deal with tenants but not be a "person managing" because they have nothing to do with the rent at all. This is even more true in relation to the various regulations which specify how an HMO must be managed.[95] These regulations use the term manager in a more plain English sense, but in regulation 2, it is clear that manager means "person managing". Occasionally, local authorities and others are led into error by mistaking an individual being described as a "manager" as meaning that they are the "person managing".

The "person having control" is any person who receives the rack-rent for the relevant premises either for themselves or for someone else by acting as their agent or trustee or who would receive the rack-rent if the premises were let at a rack-rent. A rack-rent is defined[96] as a sum which is not less than two-thirds of the net annual value of the property. The net annual value is an old definition used for property rating, and in those cases, it is the value that a property could earn in rental in an open market over the course of a year. So, any person who is to receive or could receive at least two-thirds of the annual rental will be a person having control. In practice, almost all persons having control will also be persons managing due to the very wide definition of person having control. This definition is really there to capture situations where someone lets a property for no rent to seek to evade the person managing definition.

However, the idea of "rack-rent" is less than perfect. A property might have more than one plausible market rent, especially if it is an HMO. The rent for a property let as a whole will be very different to the aggregate rent of it when let on a room-by-room basis as an HMO. In fact, the difference can be great enough that a market rent for a property let as a whole might well be less than two-thirds of the aggregate of the HMO room rent. Therefore, which figure should be used when considering the rack-rent? A landlord who has let his or her property as a whole to a single tenant who has then sub-let the same property on a room-by-room basis as an HMO might easily receive less than two-thirds of the rent being received by their tenant from the actual occupiers. If the rent being received from the occupiers is seen as the base figure to calculate the rack-rent, then the superior landlord will not receive a rack-rent. However, they are still likely to be caught as they would be the person who "would" receive that rack-rent had it been let out by them for a rack-rent.

However, it is not just the rent that counts. In relation to a person managing, the legislation actually covers more than just rent. The wording of the legislation actually specifies "rents or other payments". This is likely to include payments in respect of utilities and council tax for example. It may also include a tenancy deposit. So, there has generally been a view, especially among letting agents, that they are not liable for licensing failures if they are only operating on a tenant find basis because they are not collecting the rent. However, in practice, a letting agent even on a tenant find will usually collect the first month's rent and the tenancy deposit from the tenants. So, they will have taken "rents or other payments" within the meaning of the section. However, this interpretation places agents in a potentially unfair position. If they take rent and a deposit from a landlord who has applied for a licence, then they will not have done anything wrong at the time. If the landlord is later refused a licence, then the agent would have been a person managing the relevant property and it would not be licensed leaving them potentially liable for this failure. The courts have never clarified if the definition of person managing only applies at the time the money is taken or if it lasts so that once I am a person managing, I can never escape. The latter definition, while one which would follow a strict reading of the legislation, is one which is unlikely to be accepted as it creates an inescapable long-term liability which was presumably not what Parliament intended.

The definitions of person managing and person having control have a lot of difficulties in practice. Local authorities will often get confused and prosecute someone as a person having control when they are in fact a person managing. In addition, many letting agents offer a service where they are solely responsible for the collection of rent but not for the management of the property. In practice, these agents will still be the person managing and probably the person having control as well. Conversely, and somewhat perversely, it is also possible to actually be managing a property but not be a "person managing". If an agent is responsible for the handling of repairs and dealing with the tenant but is not involved in any aspect of the collection of rent, then they will be seen, in the plain English sense, as managing the property but they will not be managing it for the purposes of the legislation.

The nature of the definition also means it reaches all the way up the chain. So, a freeholder of a property who lets it to an individual who then sub-lets it as an HMO is a person managing and a person having control as is the individual they have let it to. If the property is not licensed, then both of them will be liable for that failure. In practice, a local authority may use its discretion to prosecute to decide whether to prosecute both parties or just one of them. The UT has been clear that it is not open to one party to a chain of landlords and tenants to say that they are not in receipt of rack-rent or cannot be in receipt of it and have no liability for licensing merely because someone else is in receipt of that rack-rent. If the tests under the 2004 Act are made out, then liability will be found even if another party could be also, or is also, liable. Even if a party could not obtain a licence at all as they do not meet the standards of the local authority that will not allow them to escape responsibility. The UT has further said that it is bad planning on the part of a landlord to place themselves in a position where they are liable to obtain a licence that they cannot in fact obtain and that they cannot escape liability for licensing based on it.[97]

This may sound harsh but it is ultimately based on the intention of Parliament when it elected to make HMO offences ones of strict liability. Allowing a situation to arise in which nobody was liable or in which parties could arrange affairs between themselves to divest themselves of liability is a general principle recognised in the criminal law and would not be one that would accord with the clear intention of Parliament.[98]

Companies

The legislation is not limited to natural persons and so a company that has a legal persona, such as a limited company or limited liability partnership, can be a person managing or a person having control if it is in receipt of the rent. However, this is often an area of confusion for landlords and local authority officers. Careful thought needs to be given to who owns the property, which name is on the tenancy agreement, and which bank accounts have money within them.

It is particularly important that the correct company is proceeded against using its full name accurately. Incorrect descriptions of companies in prosecutions usually renders the prosecution invalid.[99] It is common to make errors of this sort by confusing a trading name or style with the true legal name of the company. However, it is not possible to cheat the system by seeking to change a company name once proceedings have begun to avoid the proceedings.[100] If a company is struck off the register, either because it has not complied its obligations or because the officers have asked for it to be struck off or liquidated, then any prosecution against it will collapse. Local authority officers should be alert to this happening and if they become aware of it, an application can be made to the High Court to prevent the striking off. Such applications are usually granted but not all officers are aware of this and keep an eye on the situation. In principle, it is possible to apply to have a company that has been struck off restored to the register of companies but this is a great deal more difficult to do.

Just because the company is a limited company that does not mean that individuals are able to avoid prosecution.[101] Anyone who is a director, manager, or company secretary of a limited company that has committed an offence under the Housing Act 2004 is also at risk of prosecution. So are members of co-operatives or partnerships as are people who claim to act in any of these capacities. So, a person who is hiding behind another person who is the named director is potentially liable as well if they can be shown to have held themselves out as a manager or director.[102] Where it can be shown that such a person was aware of the breach or that they were negligent as to whether the breach was occurring or not, then they can be charged. In similar legislation relating to Health and Safety,[103] it has been held that there is no actual need to prosecute the company at all provided that it can be shown that the company has committed the offence. This applies even where a company has been struck off. It is not clear whether the same principle can be applied to Housing Act 2004 matters but it is likely that it can. However, just because a person is involved with the running of a company that does not mean that they are automatically responsible for its failings. Individuals are only liable if the offences committed by the company were carried out with their "consent or connivance" or as a result of "any neglect on the part of" the individual.[104] In practice, it can be difficult to show the necessary level of intent to be confident of a prosecution against an individual.

Temporary Exemption Notices

There is an alternative to licensing of a sort. This is the Temporary Exemption Notice or TEN (Figure 3.2).[105] This allows for a short-term exemption to licensing, whether mandatory, additional, or selective in limited circumstances. It is a three-month exemption only and is only for cases where the landlord is taking reasonable steps to procure that the property will not require licensing. Any application for a TEN would need to give reasons why the property has not been licensed and show what is being done to resolve the situation. It would not normally be good enough to say that there had been ignorance of licensing. A TEN is more intended for situations where a property has become licensable due to tenant misuse or where a person has found themselves unexpectedly in possession of such a property as a result of a repossession or inheritance. An extension to the original three-month TEN for a second three months is possible. However, it is even more limited in scope and it is described in the legislation as only being permitted where there are "exceptional circumstances".[106] This is a rare set of occurrences and a landlord should not seek a temporary exemption with the presumption that the initial three months will be easily extended for a further three months. At the very least, a landlord seeking an extension would need to show that they had been active in progressing the situation in the first three months and that they were well on the way to achieving a situation where licensing would no longer be needed.

Some local authorities require a formal written application to be made either using the same form as a licence application or in a very similar style. It is debatable whether this is in fact permitted. There is nothing in the legislation which allows a local authority to specify the form of an application. It merely states that a notification must be sent and the local authority should grant the exemption if it considers it fit to do so. It is more likely that the correct position is that a local authority should not have a set form but should give guidance on what sort of information it is likely to require to be able to grant an exemption and then deal with each case as it arises.

Temporary Exemption Notice (s62) Process Flow

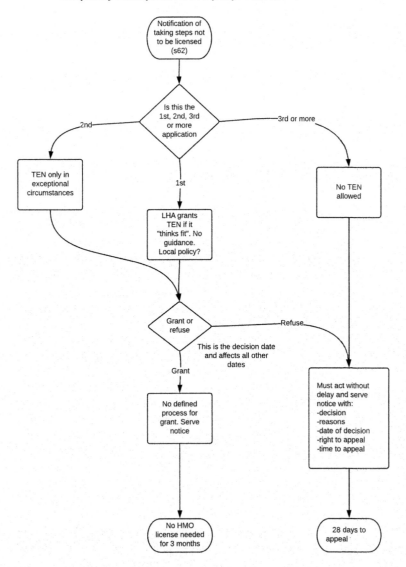

Figure 3.2 Temporary Exemption Notice (s.62) process flow.

If a local authority elects to grant a TEN, then they should provide a notice specifying that the exemption has been granted.[107] If they decline to grant the exemption, then they are required to provide a written notice of refusal "without delay".[108] That refusal must specify:

- the decision made;
- the reasons for the decision and the date of it;
- the right to appeal against the decision; and
- the time limit for that appeal which is 28 days from the date of decision.

A refusal to grant a TEN can be appealed to the FTT (or the Welsh RPT in Wales) and this appeal must be lodged within 28 days of the date the decision was made. The Tribunal will then consider the application afresh and either keep the decision of the local authority or substitute their own decision. In practice, if the deadline is missed, there is no limit on simply making a further application for a TEN and then appealing the further refusal.

Revoking Licences and Termination

Normally, licences continue in force until the date of expiry that was set out when they were first granted or until they are revoked (Figure 3.3).

There are also other scenarios which allow for a licence to end but the law in this area is somewhat confused. The legislation is clear that a licence must not have a duration longer than five years[109] and also specifies that it continues in force until such time as its set duration expires, it is expressly revoked, or it ends as a result of the death of the licence holder.[110] Where a licence is revoked on death, there is also a special exemption that exists which permits the property to be treated as licensed for a period of three months to allow for a new licence holder to come forward.[111] However, the UT has also held that where a property is sold, the licence is automatically revoked stating that it

> remains the law that where a property is sold and the new owner takes over management and control from the seller, that new owner requires a licence.[112] The previous licence cannot be transferred to the new owner and is of no assistance, whether or not expressly revoked, because the new owner does not have a licence.

In summary therefore, the UT analysis was that it was up to the new owner to have a licence and because the licence could not be transferred the new owner did not have a licence and therefore the property was unlicensed. This is a controversial decision which has come under criticism.[113] The problem with the approach taken by the UT is that it opens up a new version of the so-called registration gap that has plagued property notices.[114] In short, a property sale does not happen in one neat step. Contracts for sale are exchanged and completed but

66 *Licensing of Properties*

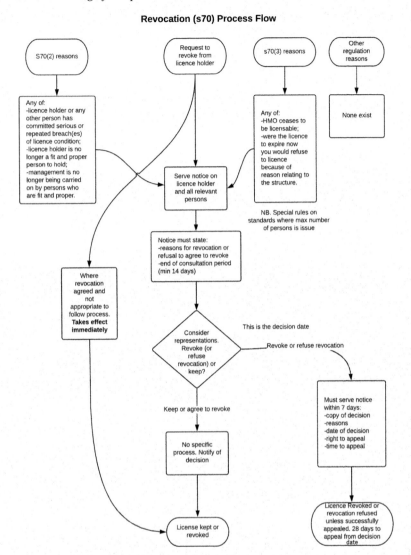

Figure 3.3 Revocation process flow.

the legal title does not transfer until an application has been made to the Land Registry and it has processed the transfer and amended the register entry. The register entry is then backdated to the date that the application was made to the Land Registry. The UT decision does not sit well with this process. Is the licence deemed to be revoked when contracts are exchanged or does it happen when the register is updated? Given that the register is backdated, the second option would mean that the new owner was then breaking the law as the licence would have expired at the point of the backdated register update.

There is also a process which allows for formal revocation of the licence. This can be with the agreement of the person who holds it and the sub-text of the legislation is that this would normally be at their request.[115] In this case, the revocation takes place immediately on the local authority making the decision to revoke.[116] Otherwise, the local authority can revoke a licence where it has been breached in a serious manner, where the licence holder is no longer fit and proper, or where the local authority considers that the management is not being carried out by people who are fit and proper.[117]

There is a further power to revoke relating to the condition of a property, but the wording of this part of the legislation and the manner in which it should operate is unclear. One reading of the wording is that a revocation can occur where a property has stopped being an HMO, but the licence is ongoing, and the local authority then considers that were an application for a licence made at that point they would refuse it because of reasons relating to the structure of the HMO. The alternative reading is that these are two entirely separate grounds for revocation and so a local authority can revoke a licence either when a property has ceased to be an HMO or when it considers it would not grant a licence if an application was made to it. The second reading is a more natural way of reading the legislation from a linguistic point of view. It also makes sense that a local authority might wish to end a licence that was effectively not being used. However, if a local authority is able to revoke a licence on the basis that they are unhappy with the structure, then the property must previously have been granted a licence and so it must have been deemed suitable at some stage. In addition, if the property is currently in use as an HMO, then a revocation would mean that the landlord would immediately be operating an unlicensed HMO and would be committing an offence. This is a very hard reading of the legislation and it seems unlikely that Parliament intended to place landlords in such a difficult position.

When considering whether it would refuse to grant a licence on fresh application, the local authority must do so on the basis of the standards that existed at the time the original licence was granted unless there has been a change to a national standard set by central government in which case they should apply that new national standard.[118] This is clearly intended to allow for a local authority to revoke a licence between tenancies if they become concerned about the physical property or if national government changes the standards applicable. There is also a power to add further criteria to these revocation criteria by way of a statutory instrument, however none has been made.

In England only, there is also an absolute duty on a local authority to revoke a licence where a person who holds the licence is banned or where a person owning an estate or interest in the property and acting as a landlord or licensor is banned.[119]

Where a local authority has decided to revoke or refused to revoke a licence, they must serve a notice on:

- the licence holder;
- any person having control or person managing the property; and
- any person with an estate or interest in the property save for tenants with less than three years remaining on their tenancy.

There is something of a flaw in this structure in that for most tenants on standard Assured Shorthold Tenancies there is no obligation to inform them that the property they are living in has had its licence revoked. The licensing status of a property is of particular importance to an Assured Shorthold tenant as they cannot be served with a notice seeking possession under s.21, Housing Act 1988 where a property is unlicensed or does not have a licence application before the local authority. Therefore, not being told that a licence has been refused or revoked would leave them potentially at the mercy of an unscrupulous landlord. Ideally, a local authority would elect to simply send a notification to the tenants regardless in order to make sure that they were made aware of the situation.

A revocation notice must specify the reasons for the revocation and the rights of appeal against the decision, including time limits.[120] Where the revocation is being made by agreement, or in England pursuant to a banning order, that is all that is needed. If the revocation is to be made for some other reason and in all cases of refusal to revoke, then the local authority must first serve a notice setting out its intention to revoke or refuse to revoke a licence. This notice must be served on the same group of people and set out the reasons for the decision to revoke or to refuse to revoke and invite representations before a date not less than 14 days in the future.[121] Those representations must be considered before a final decision is made.

In relation to selectively licensed properties, a similar series of powers exist. The process and reasons for revocation as a result of persons involved no longer being fit and proper[122] or, in England, where a relevant person has been banned.[123] There is also a series of powers to revoke where a property is no longer required to be licensed under selective licensing scheme, where an HMO licence has been granted instead, or in a similar manner to HMO licensed property where the property is such that if a licence was sought today it would be refused due to some aspect of its structure.[124]

While there is no right of appeal in relation to revocations made with the agreement of the licence holder or pursuant to a banning order being made, the notice specifying that a decision to revoke has been made must still be served.[125]

Appeals are to the FTT or the Welsh RPT, as appropriate, who will consider the entire matter afresh. They can confirm, reverse, or vary the decision of the local authority as they see fit.

Varying Licences

There is also a general power to vary a licence[126] and a procedure for doing so (Figure 3.4).[127] A variation can be made by the local authority acting itself or at the request of any of:

- The licence holder;
- Any person managing or having control of the property; and
- Any person with an estate or interest in the property.

However, a tenant who has a tenancy of less than three years cannot make such a request.[128] As with a revocation, the local authority, when considering a variation of an HMO licence, can only apply the standards that were in force at the time the licence was originally granted unless there has been a variation of the nationally approved standards in which case the local authority has the option to apply these.[129]

If a variation is requested or the local authority decides that they wish to make one, then they must serve a notice on:

- The licence holder;
- Any person managing or having control of the property; and
- Any person with an estate or interest in the property;

70 *Licensing of Properties*

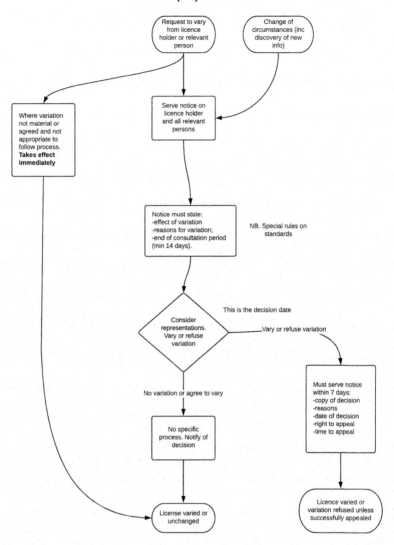

Figure 3.4 Variation (s.69) process flow.

that the local authority are aware of. There is no requirement to serve the notice on any tenant who has a tenancy of less than three years. This notice must specify:

- That the authority are proposing to make a variation;
- What the effects of the variation will be;

Licensing of Properties 71

- The reasons for the variation; and
- How long there is to make representations on this proposal which must be at least 14 days from the date of service of the notice.[130]

The local authority is not required to serve a notice if they are proposing to make a variation that they or someone else has proposed and:

- They consider that the variation is not material, or
- The variation is agreed by the licence holder and the authority feels that it is not appropriate to serve a proposal notice.[131]

The local authority can also refuse to serve a proposal to vary notice if:

- They have already served a proposal to vary notice in relation to a previous variation, and
- Consider that the variation now being proposed is not materially different from the previous proposed variation.[132]

This second set of limits is sometimes misconstrued by local authorities as permitting them to refuse to serve notices or ignore variations that they consider to be repeated when they have already been refused. This is not correct as the ability to decline to serve a notice of proposal to vary only applies in situations where the local authority is proposing to actually grant a licence.

As always, the local authority must consider representations received and then make a decision to vary the licence or to decline to do so. If they intend to refuse the variation, then there is a requirement to serve a notice of intention to refuse the variation. This needs to be served on all the same people as the notice of a proposed variation.[133] The notice must specify:

- That the authority is proposing to refuse to vary the licence;
- Their reasons for the refusal; and
- The period for which representation can be made which must be at least 14 days from the date of service of the notice.[134]

Once this additional period of consultation has occurred, then the local authority must serve a final decision notice. Or if the intention is to vary the licence, then a final decision notice must be served without the additional consultation. This final notice must again be served on all the same people as a proposal notice needs to be served on and must set out:

- The decision to vary or to refuse the variation along with the date of decision;
- The reasons for that decision;
- The right to appeal that decision; and
- The time period during which an appeal must be lodged which will be 28 days from the date of decision.[135]

As with revocations, there is a right to appeal to the FTT or Welsh RPT within 28 days of the decision. They can confirm or overturn the decision of the local authority or vary it as they see fit.

A variation comes into force immediately if it is agreed with the licence holder, but if not, it will come into force once the time limit to appeal has passed or if an appeal is lodged once those appeals are resolved and the time limit for further appeals has passed.

The power to request a variation and then appeal a refusal actually creates a bit of a loophole in the time limits for appeals against HMO licences. The reason for this is that a local authority may grant a licence with conditions in it and the licence holder or some other person may not appeal within the 28-day time period to do so. The local authority may then consider that the time to appeal has passed and they need no longer concern themselves with the situation. However, it would be open to the licence holder to apply to the local authority to have the licence varied to remove or change a condition that they found unacceptable and to then appeal the decision of the local authority on that request for a variation. So, there are aspects of licences which remain open to appeal throughout their lifetime.

Register of Licences

Local authorities are required to keep a register of all the HMO and selective licences and all temporary exemption notices. They must also keep a register of management orders that are in force.[136] The data that is required to be kept on the register is set out in regulations.[137] In relation to licences, the register must contain for each licence the following data:

- the name and address of the licence holder and the person managing the licensed property;
- the address and a short description of the licensed property;
- a summary of the conditions of the licence;
- the commencement date and duration of the licence; and
- a summary of any issue concerning the licensing of the property that has been referred to a tribunal along with any decision of a

tribunal in relation to the property, including the tribunal's case reference number.

For HMOs only, the register must also set out for each licensed HMO:

- the number of storeys in the HMO; and
- the number of self-contained flats (if any).

For HMOs that are s.254 HMOs, the entry must also include:

- the number of rooms which include:
  - sleeping accommodation; and
  - living accommodation;
- the number non-self-contained flats (if any);
- a description and the numbers of any shared amenity; and
- the maximum number of persons or households permitted to occupy the HMO under the conditions of the licence.

For temporary exemption notices, the register must contain:

- the name and address of the person who has sought a TEN;
- the address of the property the TEN applies to and any reference number applicable;
- a summary of the effect of the TEN;
- details of any previous TENS for the property;
- a statement of what the person requesting the TEN intends to do to procure that the property will cease to be licensable;
- the date on which the authority served the TEN and the date it comes into force; and
- a summary of any issue concerning the TEN that has been referred to a tribunal along with any decision of a tribunal in relation to the property, including the tribunal's case reference number.

In respect of management orders, the register must contain:

- the address of the HMO or house to which the order relates and any reference number allocated to it by the local housing authority;
- a short description of the HMO or house;
- the date on which the order comes into force;
- a summary of the reasons for making the order;
- a summary of the terms and type of order made; and

- a summary of any issue concerning the TEN that has been referred to a tribunal along with any decision of a tribunal in relation to the property, including the tribunal's case reference number.

Where the management order is connected to an HMO, then the register must also specify:

- the number of storeys in the HMO; and
- the number of self-contained flats (if any).

For HMOs that are s.254 HMOs, the entry must also include:

- the number of rooms which include:
  - sleeping accommodation; and
  - living accommodation;
- the number non-self-contained flats (if any);
- a description and the numbers of any shared amenity; and
- the maximum number of persons or households permitted to occupy the HMO.

The issue of how the register is managed is one which is contentious. Some local authorities simply place the entire register on a website and allow anyone to access it. While there is no set rule about how the register must be stored and the format it must be kept in, to have the entire register accessible in this way is unlikely to meet a local authority's obligations to protect personal data.[138] A register will contain personal data of at least some people who have obtained licences or sought TENs and to publish all that data indiscriminately without any consideration of the reasons why somebody may wish to access it is unlikely to be proportionate or reasonable. A local authority must make the register available for inspection at its offices and may charge a fee for supply of part or all of the register. However, this would permit a more balanced approach which would allow reasonable investigation to ensure that any person accessing the register wanted to use it for an appropriate purpose consistent with the requirement to properly protect personal data.

Codes of Practice

There is a power for the SoS or the Welsh Ministers to prescribe a national code of practice for the management of HMOs or in relation

Licensing of Properties 75

to any of the exemptions from a property being an HMO.[139] That code can be different for different types of HMO or different types of HMO landlord and can be different in different geographical areas. There is only one such code approved at this stage. That is for properties managed by recognised educational establishments which are exempt from being HMOs if they are accommodating their own students and managed in accordance with the approved code.[140]

A failure to comply with a code is not something that leads to a direct penalty in itself. It may however lead to removal from some other scheme or an inference that the person has breached some other aspect of the law relating to the management of HMOs.

Notes

1. Licensing of Houses in Multiple Occupation (Prescribed Description) (Wales) Order 2006. A similar problem existed with the, now repealed, Licensing of Houses in Multiple Occupation (Prescribed Descriptions) (England) Order 2006.
2. In *London Borough of Islington v The Unite Group Plc* [2013] EWHC 508 (Admin).
3. *Bristol City Council v Digs (Bristol) Ltd* [2014] EWHC 869 (Admin).
4. In the Licensing of Houses in Multiple Occupation (Prescribed Descriptions) (England) Order 2006 and the Licensing of Houses in Multiple Occupation (Prescribed Descriptions) (Wales) Order 2006.
5. In the Licensing of Houses in Multiple Occupation (Prescribed Description) (England) Order 2018.
6. Set out in the Licensing of Houses in Multiple Occupation (Prescribed Descriptions) (Wales) Order 2006.
7. Under the Housing Act 2004 (Additional HMO Licensing) (Wales) General Approval 2007.
8. Under the Housing Act 2004: Licensing of Houses in Multiple Occupation and Selective Licensing of Other Residential Accommodation (England) General Approval 2010.
9. The Housing Act 2004: Licensing of Houses in Multiple Occupation and Selective Licensing of Other Residential Accommodation (England) General Approval 2015.
10. *R v Home Secretary, ex parte Doody* [1994] 1 AC 531.
11. *R(Peat) v Hyndburn BC* [2011] EWHC 1739 (Admin).
12. As approved by the UK Supreme Court in *R (Moseley) v London Borough of Haringey* [2014] UKSC 56.
13. See *R (Regas) v London Borough of Enfield* [2014] EWHC 4173 (Admin), para 47.
14. See *R (Regas)*, op cit, para 48.
15. Section 56(3)(b) and 80(9)(b).
16. *R (Moseley)*, op cit.
17. Sections 56(3)(a) and 80(9)(a), Housing Act 2004.
18. See *R (Regas)*, op cit.

19 Government Legal Department (2016) *The Judge over Your Shoulder – A Guide to Good Decision Making*. Unknown place of publication: Government Legal Department. [Online] [Accessed on 21 March 2021] https://www.gov.uk/government/publications/judge-over-your-shoulder.
20 Sections 60(2) and 84(2), Housing Act 2004.
21 In sections 59 and 83 for Additional HMO and selective licensing respectively.
22 In regulation 9 of the Licensing and Management of Houses in Multiple Occupation and Other Houses (Miscellaneous Provisions) (England) Regulations 2006 and the Licensing and Management of Houses in Multiple Occupation and Other Houses (Miscellaneous Provisions) (Wales) Regulations 2006.
23 In *Thanet DC v Grant* [2015] 10 WLUK 800 (DC).
24 Under the general duty to take all steps to secure that licence applications are made, found in s85(4).
25 Sections 60(3) and 84(3), Housing Act 2004.
26 Sections 60(4) and 84(4), Housing Act 2004.
27 Sections 60(5) and 84(5), Housing Act 2004.
28 Sections 60(6) and 84(6), Housing Act 2004.
29 In regulation 10 of the Licensing and Management of Houses in Multiple Occupation and Other Houses (Miscellaneous Provisions) (England) Regulations 2006 and the Licensing and Management of Houses in Multiple Occupation and Other Houses (Miscellaneous Provisions) (Wales) Regulations 2006.
30 https://democracy.towerhamlets.gov.uk/ieDecisionDetails.aspx?AIId=124131 [last accessed on 06 March 2022].
31 See s55(5) for HMO licensing and s79(5) for selective licensing.
32 *British Oxygen Co Ltd v Board of Trade* [1971] AC 610, [1970] 3 All ER 165, HL.
33 *Metropolitan Police Commissioner,* ex parte *Blackburn* [1968] 2 QB 118.
34 Section 63(2) and s87(2), Housing Act 2004.
35 The Licensing and Management of Houses in Multiple Occupation and Other Houses (Miscellaneous Provisions) (England) Regulations 2006 and the Licensing and Management of Houses in Multiple Occupation and Other Houses (Miscellaneous Provisions) (Wales) Regulations 2006 in Wales.
36 *R (Gaskin) v Richmond Upon Thames London Borough Council* [2017] EWHC 3234 (Admin).
37 As a result of changes made by the Licensing and Management of Houses in Multiple Occupation and Other Houses (Miscellaneous Provisions) (Amendment) (England) Regulations 2012.
38 See s72(4)(b) and s95(3)(b), Housing Act 2004.
39 *R (Gaskin)* [2017] op cit.
40 *Church Commissioners for England v Hampshire County Council* [2014] 2 EGLR 203.
41 See, for example, *Saunders v Jordan*, First Tier Tribunal (Property Chamber), CAM/00KF/HMF/2021/0008 & 9, 13 January 2022.
42 More properly, Regulation (EU) 2016/679 of the European Parliament and of the Council of 27 April 2016 on the protection of natural persons with regard to the processing of personal data and on the free movement of such data, and repealing Directive 95/46/EC (General Data Protection Regulation).

43 GDPR, Article 6(c) and (e).
44 Section 63(7) and s87(7).
45 *Crompton v Oxford City Council* CAM/38UC/HMV/2013/0006-7.
46 *R (Gaskin) v Richmond upon Thames LBC* [2018] EWHC 1996 (Admin).
47 Directive 2006/123/EC of the European Parliament and of the Council of 12 December 2006 on services in the internal market. This is implemented in the UK by the Provision of Services Regulations 2009.
48 *R (Hemming t/a Simply Pleasure) v Westminster City Council* (C-316/15).
49 Section 72(4) and s95(3), Housing Act 2004.
50 See sections 55(5)(b) and 79(5)(b), Housing Act 2004.
51 Para 12, Schedule 2, Houses in Multiple Occupation Act (Northern Ireland) 2016.
52 As the Provision of Services Regulations 2009.
53 By the Provision of Services (Amendment etc.) (EU Exit) Regulations 2018.
54 See Regulation 19, Provision of Services Regulations 2009.
55 Regulation 4, Houses in Multiple Occupation (Notice of Application) Regulations (Northern Ireland) 2019.
56 Section 64(3), Housing Act 2004.
57 Section 88(3), Housing Act 2004.
58 Section 64(4), Housing Act 2004.
59 In the Licensing and Management of Houses in Multiple Occupation and Other Houses (Miscellaneous Provisions) (England) Regulations 200 and the Licensing and Management of Houses in Multiple Occupation and Other Houses (Miscellaneous Provisions) (Wales) Regulations 2006.
60 By the Licensing and Management of Houses in Multiple Occupation (Additional Provisions) (England) Regulations 2007 and the Licensing and Management of Houses in Multiple Occupation (Additional Provisions) (Wales) Regulations 2007.
61 By regulations 2 and 5 in the Houses in Multiple Occupation (Certain Converted Blocks of Flats) (Modifications to the Housing Act 2004 and Transitional Provisions for section 257 HMOs) (England) Regulations 2007 and the Houses in Multiple Occupation (Certain Blocks of Flats) (Modifications to the Housing Act 2004 and Transitional Provisions for section 257 HMOs) (Wales) Regulations 2007.
62 *Clark v Manchester City Council* [2015] UKUT 0129 (LC).
63 Sections 66 and 89.
64 Section 5, Rehabilitation of Offenders Act 1974.
65 Section 4, Rehabilitation of Offenders Act 1974.
66 Section 4(6), Rehabilitation of Offenders Act 1974.
67 See *Hussain & Ors v LB Waltham Forest* [2020] EWCA Civ 1539.
68 *London Borough Of Waltham Forest V Reid* [2017] UKUT 396 (LC).
69 Sections 125(3) and 125(6), Housing and Planning Act 2016.
70 Regulation 21, Provision of Services Regulations 2009.
71 Immigration Act 2014. Requirement added by s125(1), Housing and Planning Act 2016.
72 Added by s25, Housing and Planning Act 2016.
73 *R (Gaskin)* [2017] op cit.
74 Imposed by s76, Housing Act 2004.
75 *Matheson v Cardiff CC* RPT/WAL/HMO/1.

76 Schedule 4.
77 *The Landlords Association for Northern Ireland, Re Application for Judicial Review* [2022] NIQB 14.
78 *Nottingham City Council v Parr & Anor* [2018] UKSC 51.
79 *Brown v Hyndburn Borough Council* (2018) EWCA Civ 242.
80 Section 67(1).
81 Section 90(1).
82 Section 67(1)(b).
83 Under provisions in s80(7).
84 By the Selective Licensing of Houses (Additional Conditions) (England) Order 2015 (SI 2015/977).
85 *Crompton v Oxford CC*, First Tier Tribunal (Property Chamber), CAM/38UC/HMV/2013/0006-7, 28 March 2013.
86 In the Licensing of Houses in Multiple Occupation (Mandatory Conditions of Licences) (England) Regulations 2018.
87 Office of the Deputy Prime Minister (2006) *Housing Health and Safety Rating System: Enforcement Guidance*. London: Office of the Deputy Prime Minister. [Online] [Accessed on 22 March 2022] https://www.gov.uk/government/publications/housing-health-and-safety-rating-system-enforcement-guidance-housing-conditions.
88 See, for example, *Ultimate Housing Ltd v LB Southwark*, First Tier Tribunal (Property Chamber), LON/00BE/HPO/2013/0021, 23 August 2013.
89 Paras 1–13, Schedule 5, Housing Act 2004.
90 Paras 3–4, Schedule 5.
91 Paras 14–21, Schedule 5, Housing Act 2004.
92 Para 31, Schedule 5.
93 Para 34(2), Schedule 5.
94 Para 34(4), Schedule 5.
95 Management of Houses in Multiple Occupation (England) Regulations 2006 and the Management of Houses in Multiple Occupation (Wales) Regulations 2006.
96 In section 263(2).
97 *Urban Lettings (London) Ltd v LB Haringey* [2015] UKUT 104 (LC).
98 *Pollway Nominees v Croydon LBC* [1987] AC 79.
99 *R (J Sainsbury plc) v Plymouth Magistrates' Court* (2006) 170 JP 690.
100 Companies Act 2006, s81.
101 Under s251, Housing Act 2004.
102 See 251(1)(b), Housing Act 2004.
103 Health and Safety at Work Act, s37.
104 Section 251(1), Housing Act 2004.
105 Allowed for by s62 and s86, Housing Act 2004.
106 Sections 62(5) and 86(5).
107 Sections 62(2) and 86(2).
108 Sections 62(6) and 86(6).
109 Housing Act 2004, s68(4).
110 Housing Act 2004, s68(3)(b).
111 Housing Act 2004, s68(8).
112 *Taylor v Mina an Ltd* [2019] UKUT 249 (LC).
113 Smith D (2020) "Papering Over the Cracks in Property Licensing", *JHL*, no 2: 28–33.

114 *Baker & Anor v Craggs* (Rev 1) [2018] EWCA Civ 1126.
115 Section 70(1)(a), Housing Act 2004.
116 Section 70(7), Housing Act 2004.
117 Section 70(2), Housing Act 2004.
118 Section 70(3), Housing Act 2004.
119 Section 70A, Housing Act 2004.
120 Paragraph 24, Schedule 5, Housing Act 2004.
121 Paragraph, 22, Schedule 5, Housing Act 2004.
122 Section 93(2), Housing Act 2004.
123 Section 93A, Housing Act 2004.
124 Section 93(3), Housing Act 2004.
125 Para 32(2) and 32A, Schedule 5.
126 Sections 69 and 92, Housing Act 2004.
127 In paras 14–21, Schedule 5, Housing Act 2004.
128 See s69(8) and s92(5).
129 See s69(3) and s69(4).
130 Para 14, Schedule 5, Housing Act 2004.
131 Para 17, Schedule 5.
132 Para 18, Schedule 5.
133 Para 19, Schedule 5.
134 Para 20, Schedule 5.
135 Paras 16 and 21, Schedule 5.
136 S232, Housing Act 2004.
137 In regulation 13 of the Licensing and Management of Houses in Multiple Occupation and Other Houses (Miscellaneous Provisions) (England) Regulations 2006 and regulation 11 of the Licensing and Management of Houses in Multiple Occupation and Other Houses (Miscellaneous Provisions) (Wales) Regulations 2006. Amended by the Licensing and Management of Houses in Multiple Occupation (Additional Provisions) (England) Regulations 2007 and Licensing and Management of Houses in Multiple Occupation (Additional Provisions) (Wales) Regulations 2007, respectively.
138 Under the Data Protection Act 2018.
139 In s233, Housing Act 2004.
140 By the Housing (Approval of Codes of Management Practice) (Student Accommodation) (Wales) Order 2006 and the Housing (Codes of Management Practice) (Student Accommodation) (England) Order 2010.

4 Taking Over and Managing Properties

In specific circumstances, there are powers which enable a local authority to take over the management of a licensable property by making a management order.[1] The three scenarios are that:

1. The local authority has refused to grant a licence and they consider that there is no reasonable prospect of the property being licensed in the near future.
2. The local authority has revoked an existing licence and that when the revocation comes into effect there will be no reasonable prospect of the property then being licensed in the near future.
3. The local authority has refused to grant or revoked a licence for a property and they consider that taking over management is necessary for protecting the health, safety, or welfare of the occupiers or persons occupying or having an estate or interest in any premises in the vicinity.

In summary then, the first two situations are designed to address situations in which the proposed licence holder is not suitable and no other party who the local authority consider suitable has been or is likely to be put forward while the third is intended to deal with situations where the property is putting people at risk.

Additionally, there is a separate ability to seek a management order where there is a risk to the health, safety, or wellbeing of the occupiers or to people carrying out lawful activities in or around the property; and, in addition, the property meets the criteria set out in the relevant regulations.[2] These are that:

- the area in which the house is located is experiencing a problem with anti-social behaviour which is at least partly caused by anti-social behaviour of an occupier of the house;

- the landlord of the house is failing to take reasonable action to deal with the problem; and
- the making of an management order, combined with other measures taken in the area by the local housing authority or others, will reduce or eliminate the problem.

Interim and Final Orders

Management orders fall into two types, interim and final orders. As the names suggest, interim orders are more temporary in nature with a maximum length of 12 months while final orders are longer lasting, subject to a maximum length of five years. In general, interim orders can be made directly by a local authority without permission, although they can be appealed against, while a final order is only made by the FTT on application by the local authority. In practice, a local authority will usually make an interim order initially to combat the immediate problem and then move to make an application for a final order if an alternative solution cannot be found. Contrary to the perception of some landlords, and the behaviour of some authorities, the making of a management order should not be a permanent deprivation of a property. Indeed, during an interim order, a local authority is obligated to either revoke the order or seek to make a final management order.[3] Even, during the operation of a final order, the local authority is under an ongoing obligation to keep the order under review and consider whether it is the best available option.[4] That said the practical upshot of a management order can often be the permanent loss of the property as the owner is at a real risk of not receiving enough money to be able to pay charges secured on the property and they can then be forced to sell or have the property repossessed.

Notes

1 In Part 4, Chapter 1, Housing Act 2004.
2 Being the Housing (Interim Management Orders) (Prescribed Circumstances) (England) Order 2006 and the Housing (Interim Management Orders) (Prescribed Circumstances) (Wales) Order 2006.
3 See s106(4) and s106(5), Housing Act 2004.
4 Under s115(3), Housing Act 2004.

5 Enforcement of the Law

Prosecutions

There are a series of potential offences associated with Houses of Multiple Occupation (HMO) properties. These are:

1. Failure to hold a licence for a licensable HMO
2. Overcrowding of a licensed HMO beyond the numbers permitted by the licence
3. Failure to keep to a required property standard in a licensable HMO
4. Failure to keep to a condition on an HMO licence
5. Failure to keep to any requirement of the HMO Management Regulations
6. Obstruction of an inspection
7. Failure to respond to a notice seeking information
8. Breach of planning enforcement

The Decision to Prosecute

Prosecutors have a wide discretion when deciding whether to prosecute. The fact that there are other courses of action available to them is not relevant, even if the local authority has not carried out the necessary processes internally to bring those options into use.[1]

In some cases, a prosecutor can commit an abuse of process in proceeding with a prosecution and the magistrates court can be asked to make an order staying the prosecution indefinitely. However, this is very rare and the threshold is a very high one.[2] Common events which might give rise to this are as follows:

- A failure to follow the enforcement policy of the relevant local authority. This is a key and increasingly useful aspect of

DOI: 10.1201/9781003297796-5

the discretion. Prosecutors should follow their own policies on enforcement.
- **Promises not to prosecute.** It is actually quite rare that a such a promise is genuinely made. Any such promise would have to be very explicit indeed and situations where it is asserted that such a promise has been made are usually a matter of a potential defendant hearing what they want to. Where a promise was made conditionally, then any breach of a condition on which the promise has been made, however minor, is likely to invalidate the promise.

In some cases, prosecutors will proceed with relatively minor prosecutions even where there is a very limited quantity of evidence or a strong defence, only to withdraw at a very late stage after significant expense has been incurred to deal with the matter.

It is important that prosecutors can be seen to have followed the proper process when deciding whether to prosecute or not. However, prosecutors have a much wider discretion in when to prosecute than is commonly believed. For example, it is generally open to prosecutors to choose to not prosecute a person who gives them evidence against another person. It is also open to a prosecutor when faced with several potentially guilty parties to pick one or more of them to prosecute. In addition, it is open to the prosecutor to choose to prosecute someone in order to send a message to the public at large, *pour encourager les autres*. In all these cases, decisions should be made rationally and for proper reasons and should be documented clearly. But if that has been done, the discretion to prosecute or not is a relatively wide one. What is important is that the proper test is followed when deciding whether or not to prosecute. This is set out in the Code for Crown Prosecutors published by the Crown Prosecution Service.[3] While HMO offences are not prosecuted by the Crown, instead being dealt with by local authorities on behalf of the Crown, those authorities are still expected to abide by it. There are some key points to note in the Code:

- **Impartiality** – The Code emphasises the importance of bringing prosecutions in an impartial manner. It is not for the prosecutor to adjudge guilt, that is for the courts. A prosecutor is assessing whether a case can be made for guilt and whether that case should be put before the courts. In practice, this is a point that many local authority officers are weak on.
- **Independence** – Prosecutors are independent. This means that they should not be subject to interference by local or national politicians. It is often thought by landlords that the answer to a

prosecution is to speak their local elected representative. This is unlikely to be effective due to the independence requirement.

The Full Code Test

The key element to consider when starting or continuing a prosecution is the, so-called, Full Code Test. In practice, this is the test that every local authority should consider and satisfy before commencing a prosecution. It is also a test that must be kept in mind and should be re-considered if something important changes. The Full Code Test has two elements which must both be met:

1. **Evidential Test** – The evidential test requires that a prosecutor must be satisfied that there is sufficient evidence to prosecute the offence. In practice, most HMO offences meet this test easily as they are state of affairs offences and there is little need to consider what was intended. However, this test should include consideration of the quality, reliability, and credibility of the evidence and whether it is admissible in court. Local authorities do sometimes make mistakes by seeking to rely on evidence that has been obtained by hearsay rather than directly.
2. **Public Interest Test** – This test is considering if the public interest is being served by prosecuting in the specific case. Many landlords will assert that it is not being served because they did not know about their responsibilities or that they made a mistake. This is not always a good argument (see more below). There is a general public interest in enforcing the law and in encouraging landlords to be wary of and comply with their obligations. This facet alone will justify most prosecutions.

State of Affairs Offences and Mens Rea

It is important to remember that most HMO offences are what are called "state of affairs offences". This means that if a particular event occurs, then the offence is established. So, if a property is licensed for five people and there are six people in it, then the offence has been committed. It is not relevant whether the person meant to do it, knew that an offence was being committed[4] or that they are really sorry. There is no requirement of intent to commit the offence as there is with a large number of offences. This concept, known as *mens rea*, is assumed by many people to apply but does not do so in these cases. The proper place for considering whether there was a sufficient level of intent is as part of the public interest test (is it appropriate to prosecute someone who has made a mistake) or in sentencing (should the sentence be reduced as the person did not intend to commit the offence).

The intention to carry out the offence is a key part of the public interest test. It is not in the public interest to use public funds to prosecute people who have made a genuine error. However, this needs to be considered carefully. Landlords have a responsibility to comply with the law that applies to their business, just like restaurants, petrol stations, or any number of other businesses. Just as with those other businesses, failure to comply, however unintended, can have serious consequences. So, it is not a strong argument for a landlord to assert that they were unaware of the HMO legislation and that is why they did not have a licence or that they did not read the licence properly and that is why they are in breach of its conditions. These are basic obligations. Accordingly, local prosecutors often apply a low weighting to intent when making a prosecutorial decision on the basis that they see most of the unintended failures as the result of carelessness.

Courts, however, do consider these issues more substantially when sentencing.

Room Sizes and Public Interest

In relation to HMO licensing, there are some problems with the test for the number of persons and the issue of prosecution. For example, a landlord may let a property to four people so that it falls outside of the mandatory licensing requirement but one of the occupiers might then fall pregnant and have a child, all without the knowledge of the landlord. Likewise, a landlord may have a licence for five people and end up having six people in the property due to a tenant giving birth. Even if the landlord became aware of the situation, it would still be impossible for him or her to do anything about it as it would constitute unlawful discrimination to evict a woman during her pregnancy or for the 26 weeks after she has given birth on the basis of her pregnancy or something directly arising from it.[5] Therefore, a landlord who evicted a pregnant woman or one who had given birth in order to prevent or resolve such a problem would be committing an offence. However, allowing the property to become an unlicensed or overcrowded HMO and not resolving the issue is also an offence, so landlords are placed in an impossible position. It is these sorts of issues that the discretion not to prosecute in the public interest was granted to prosecutors to resolve. In England, guidance has been issued indicating that central government would consider prosecutions in these cases not to be in the public interest other than in exceptional circumstances where a landlord had actively conspired to engineer the situation.[6] No similar guidance exists in Wales, but the public interest test still applies and should be considered there.

Time Limits

Prosecutions relating to HMO licensing matters are summary offences and so occur in the magistrates court. These prosecutions are subject to a statutory time limit. This specifies that an information cannot be laid before a magistrate to begin a prosecution more than six months after the offence was committed.[7] However, where the offence is ongoing, for example a landlord who has not applied for an HMO licence and has continued to fail to submit a valid application, the time period is counted from when that offence ceases to be committed. Therefore, where an offence is committed, it is important from the perspective of a landlord to file a valid application as soon as possible so that the time limit comes into effect.

In some cases, local authorities are slow to begin prosecutions and may not in fact meet the six-month time limit. If they fail to do so, then this is not a mistake that can be recovered from. In other cases, a local authority might not have entirely made up its mind to prosecute but might rush the papers before the magistrates within the six months and then send them to the suspected offender a month or two later once they have had the chance to review the evidence. This is not permitted and is an abuse of the process of the court. Prosecutors must make a decision to charge within the statutory time limit and then act on that and not seek to game the system by laying information on a range of cases speculatively and then discontinuing those that they do not think are strong.[8]

Where an offence can be tried before a magistrates or crown court, an either way offence, then there is no time limit on beginning a prosecution. Offences relating to planning, such as failing to comply with a planning enforcement notice,[9] fall into this category. In such cases, excessive delay can still prevent a prosecution but there is no set time period and it would need to be a very substantial delay indeed which had no real justification on the part of the prosecutor and would also have to create a very serious degree of unfairness to the defendant.[10]

Defences

There are some defences set out in statute to the various offences. There is a statutory defence to the offence of not having a licence that an application for one had been made or that a valid application for Temporary Exemption Notice had been made to the relevant local authority.[11]

It is also possible to advance a defence of reasonable excuse in relation to any HMO or selective licensing-related offence. However, a reasonable excuse defence is not a catch-all for any excuse. It requires a very good excuse indeed for the situation. This has led to some criticism as Parliament has placed a "reasonable excuse" defence into a range of legislation without providing any guidance or detail as what might be reasonable. This has left the courts facing a substantial burden in trying to decide what might be reasonable. In their desire not to go too far and risk shading over from interpreting to making the law, the courts have tended towards a very narrow view of "reasonable excuse".[12]

One area in which an excuse might be reasonable would be to show that the local authority themselves are in some way responsible for the situation, either because they gave bad advice which the landlord then reasonably followed or because they mis-advised the occupiers which exacerbated the situation.[13] However, even in these cases, it would be necessary to show that the actions of the local authority were a very direct cause of the offending and that it was reasonable to treat their advice as accurate.

This creates a problem for many landlords who might have rented their property to a tenant who then misuses it as an HMO without their knowledge or who have relied on an agent to obtain a licence who has failed to do so. At least in relation to Rent Repayment Orders (RROs) the Upper Tribunal (UT) has been quite cool on reasonable excuse defences where the landlord is blaming the agent.[14]

Beyond these areas, the only other defences are to show that the property is not licensable or falls outside the definitions of an HMO. These are usually very factual situations which rely on showing that the construction or use of a property falls outside the definition of an HMO or outside the licensing designation made by the relevant local authority.

Mitigation

It is often difficult to appreciate the crucial difference between a defence and mitigation. A defence, such as reasonable excuse, means a very good reason why, for example, a licence has not been obtained. This reason would have to be something that was not entirely within the contemplation of lawmakers and which justifies the allowance of the restrictive "reasonable excuse" defence.

An assertion that someone else was going to get the licence, that somebody was unwell or on holiday, or that people forgot is not enough. It is important to remember that most HMO licensing offences are state of affairs offences and so the intention to commit the offence is of little relevance. Therefore, a reasonable excuse must be something that goes beyond a mere lack of intent to something that made the commission of the offence inevitable through no fault of the Defendant.

However, many failed attempts to provide a reasonable excuse are examples of mitigation which will be relevant when considering the appropriate sentence. So, a lack of intent to commit the offence, illness, misunderstanding, or other unexpected events are all matters that a court should take into account when considering the level of sentence that they will apply to a specific offence.

Fines

As originally drafted, the maximum fine under the Housing Act 2004 for not having an HMO licence or for overcrowding a property was £20,000 and was £5,000 for all other offences. However, this was changed from 12 March 2015, and for any offence committed after that date, the fine is unlimited.[15] There is no power to give a custodial sentence of any sort. In relation to planning breaches, that is using a property as an HMO where it does not have proper planning permission for such a use, the fine is, again as from 12 March 2015, either £1,000 or unlimited.[16] In more extreme cases where there are persistent planning breaches, it is possible for a court to make an injunction requiring the breach to cease and a failure to obey such an injunction would be a contempt of court which would then be punishable by a custodial sentence.[17]

In practice, fines in the Magistrates Courts are notoriously variable and there is little certainty as to what might be awarded. In addition, contrary to popular belief, local authorities do not receive any of the money awarded in fines in the magistrates court, this money is paid to Central Funds, that is to HM Treasury. So, local authorities only receive any amount awarded in costs by the magistrates court. These sums tend to be very low indeed. A 2014 study by the Local Government Association found that in all but one of the cases they studied the costs awarded by the magistrates were less, often far less, than the cost incurred by the local authority in pursuing the prosecution.[18] Where fines are low, then it is possible that unscrupulous landlords are prepared to offend on the basis that the profit they will make exceeds the likely costs of a fine in the event that they are caught.

The Sentencing Council provides guidance on sentencing for magistrates which, where it exists, they must refer to and make use of. However, there is no specific guidance as to the level of fines to be awarded in housing cases and so magistrate can have difficulty deciding what fine to award. It is not appropriate for magistrates to refer to other decisions in other courts as these decisions are not binding on them and they would not be using their discretion properly if they were to be limited by the decisions of other magistrates. That said, there is some general overarching guidelines offered by the Sentencing Council which are applicable in specific cases. These include guidance on seriousness, totality, and early guilty pleas.

Seriousness of offence is a factor magistrates should consider in every case when deciding sentence. In doing so, they are required to consider the "offender's culpability in committing the offence and any harm which the offence caused, was intended to cause or might foreseeably have caused".[19] The Sentencing Council guidance on seriousness provides assistance with assessing culpability and harm.[20] Culpability is based on whether the offender intended the harm to result or was entirely careless as to whether or not harm resulted. Culpability will be greater where, among other things, the offender:

- targeted vulnerable people (such as vulnerable tenants);
- carried out the offence deliberately or for financial gain (such as by avoiding costs of necessary safety equipment);
- is acting in a professional capacity (such as a professional agent);
- sought to conceal evidence; or
- where the offending is part of a pattern of repeated or consistent behaviour.

Harm is increased, among other things, where there is:

- harm caused to those in public service or providing service to the public;
- multiple victims (common in HMO property); or
- a presence of others such as children or relatives of the victim.

These are not the only relevant factors but are the most relevant to offences involving housing. It is the combination of culpability and harm that allows the assessment of seriousness.

It is common in HMO and housing matters for offenders to be charged with multiple offences, sometimes as many as 20 or more. It is in these cases that the concept of totality is relevant. Where there are multiple offences, each of them will carry a fine. Fines are, by their very nature, cumulative, and so when sentencing for multiple offence, it is possible that the overall fine may become excessive. It is the principle of totality that seeks to avoid this outcome and ensure that the total fine imposed is proportionate to the offending.[21] Where two or more offences have arisen from the same incident or are committed against the same victims, the correct approach is to sentence for the most serious offence but to make the fine high enough to take into account the sum total of the offences involved. This would only not be done where the offence being sentenced had a maximum penalty that was insufficiently high to encompass all the offending. However, this should not happen in licensing matters as most of the offences involved have unlimited fines. Where the offences being sentenced arise from different or unconnected incidents, then a separate fine should be imposed for each one. Once this has been done, then the total arrived at should be considered and its overall proportionality assessed as against the offending. If the total is excessive, then the original fines should be re-assessed to reduce the total to a more appropriate figure before the fines for each offence are finalised.

There is also guidance in respect of early guilty pleas.[22] This guideline allows for a reduction that operates on a sliding scale, starting at a maximum of one-third for a guilty plea at the initial plea hearing and reduces to a low of one-tenth for a plea on the day of trial. In principle, what should occur is that the sentence should be stated by the magistrates, then they should state what reduction they are applying, then it should be applied and a final sentence declared. In practice, the reduction tends to be pretty notional. Most magistrates appear to have a figure that they consider appropriate and they take the reduction for plea into account in assessing that. In other words, the figure that is initially put forward as the sentence is uplifted by the amount that will then be deducted for the early plea so that the final figure then accords with the number that the magistrates originally considered appropriate.

The Sentencing Council also now has overarching guideline to deal with all those offences, including housing offences, that do not have any specific guideline associated with them. This may help to create more consistency in sentencing although the guideline is very general as it is intended to cover a broad range of offences.[23] The guideline gives assistance in calculating how serious the offence is by first establishing the level of culpability of the offender and then assessing the harm caused by the offending before then applying adjustment for aggravating and mitigating factors.

Civil Penalties

In England only, the law was developed to address the complaint made by local authorities that penalties in the magistrates court were too low and that costs were not awarded. This was by creating a new regime which permitted local authorities to levy direct penalties themselves in relation to certain offences. This is better for local authorities because this money is retained directly by the local authority rather than being passed to the Treasury as it is with criminal fines.

Where a local authority issues a civil penalty, the threshold is in fact higher than if they elect to prosecute. The local authority must be satisfied to the criminal standard of proof. In other words, they must be satisfied beyond reasonable doubt that the offence has been committed. This is something of an odd position to be in, in that the local authority is ultimately acting both as prosecutor and as judge. This is an area that many local authorities do not get right and there is something of a perception that civil penalties are an easier route to penalising landlords. In fact, they are harder in some ways as a local authority must be satisfied that the law has been broken before the penalty is issued. That means that they must also be reasonably satisfied that no defence to that crime exists.

Landlords can appeal a civil penalty to the First-tier Tribunal (FTT). In that case, the FTT must itself be satisfied to the same criminal standard of proof as the local authority if the civil penalty is to be upheld. However, the UT has been clear that the requirement of being satisfied beyond reasonable doubt does not require the local authority or the FTT to be satisfied beyond any doubt. Reasonable inferences can be drawn from behaviour and documents.[24]

When reviewing a penalty the UT has been clear that the FTT must have regard to the local authorities' own policy on setting penalties and how it has applied it. The FTT is not required to follow that policy and is not merely there to check the local authorities' homework but if it is to depart from the policy then it must do so in a clear and principled way and must give its reasons for doing so.[25]

In its white paper entitled "A Fairer Private Rented Sector"[26] the government suggested it would look to increase the level of fines for Civil Penalties in England and also introduce a national fining framework to create more consistency in the setting of fines.

Other Penalties

As well as direct criminal prosecutions, there are further penalties for operating an unlicensed HMO. While an HMO is unlicensed and there is no licence application for it outstanding for a decision, the landlord is not able to serve a notice under section 21 of the Housing Act 1988. There is a flaw with this process, however, in that merely making an application for a licence, even one which is doomed to failure, allows the service of a section 21 notice and the notice can continue to be served during the appeal period once the licence has been refused.

Rent Repayment Orders

As a further penalty for the operation of an HMO or selectively licensed property without a licence, there is the power for tenants and the local authority to recover money paid to a landlord. This is intended to operate in a similar manner to the Proceeds of Crime Act in order to remove any profit element obtained by a landlord as a result of committing a licensing offence. This is especially important as it has been held that the Proceeds of Crime Act does not apply in respect of rent received for an unlicensed HMO because the property could have been let regardless of the existence of a licence and it is not the letting that is the offence, but the failure to have a licence. Accordingly, the rent received cannot be the proceed of a crime.[27] Notably, however, courts have started to distinguish *Sumal* and it may be the case that money earned by overcrowding a property in excess of the numbers permitted on a licence or the savings made by not providing the proper amenities or safety systems in an HMO may be the proceeds of a crime.

Landlords or agents operating unlicensed properties can be prosecuted by the local authority and fined in the magistrates' court. RROs are additional sanctions for landlords. They can be applied for by both a local authority to recover housing benefit (or universal credit) paid in respect of the property as well as by a tenant to recover rent (or top-up to benefit payments) that they have paid.

There are now two regimes in operation as England has radically revised the RRO process.[28]

Local authorities applying for an RRO need to demonstrate that at any time during the preceding 12 months the landlord has committed an offence of not having a licence and housing benefit has been paid to the landlord during that period. The landlord does not need to have been charged or convicted of an offence, but the local authority has to prove, beyond reasonable doubt, that the landlord has committed an offence of failing to have a licence. The local authority is required to serve a notice of intended proceedings giving the landlord at least 28 days to make representations before applying to the tribunal for an RRO.

The procedure is different for tenants who wish to apply for an RRO but diverges in England and Wales. The most significant difference is that in Wales tenants are required to prove that their landlord has been convicted of an offence of failing to have a licence or that a rent repayment order has already been made in respect of housing benefit, i.e. the local authority has already made a successful application for an RRO, whereas in England, this is not required. In other words, in England, a conviction or successful RRO from a local authority is not required for a tenant to make an application for an RRO, and they can make the application on the commission of the offence only. They will need to prove the offence and so this will operate in a manner similar to a private prosecution.

The other key difference between England and Wales is that in Wales the RRO regime is only available to deal with landlords who have committed or been convicted of the offence of failing to have an HMO or selective licence, whereas in England the regime has been extended to include the full gamut of offences under the Housing Act 2004. This includes breaches of licence conditions, breaches of the management regulations, breaches of statutory notices under the Housing Health and Safety Rating System as well as unlawful eviction or harassment under the Protection from Eviction Act 1977, and using violence to secure entry under the Criminal Law Act 1977.

Tenants are not required to serve notice on the landlord and can apply straight to the Tribunal, but in Wales, they are dependent on the local authority taking action against the landlord first.

Tenants must be able to show that they paid rent during the period in which the landlord was committing the offence and they must make their application within 12 months from the date of the conviction, the local authority's RRO for the return of housing benefit, or (in England) the commission of the offence, whichever date is later. This is a scenario where the FTT is operating as an enforcer of a penalty and so the tenant must be able to demonstrate the offending to the criminal standard of proof. While that does not prevent the Tribunal drawing reasonable inferences as it does not have to achieve "cast-iron certainty"[29] but that does not oblige the FTT to draw inferences that benefit the tenant where there is a shortage of evidence or the evidence is inconsistent.[30]

Tenants are further restricted by the requirement that they can only reclaim rent paid in the 12 months prior to the date of their application to the Tribunal. This means that if the tenants move out and the local authority takes over 12 months to obtain a conviction or an RRO, then the tenants will not recover anything if they wait until after that conviction has been secured.

Once the Tribunal is satisfied that the local authority or tenant is entitled to an RRO, it must decide how much money is to be repaid. Again, the rules differ depending on the party making the application, the location of the application, and the circumstances.

In England, where the local authority is seeking an RRO for any of the set offences or the tenant is seeking an RRO for an offence other than a failure to have an HMO or selective licence and the landlord has been convicted of or received a fixed penalty notice for that offence, then the Tribunal must order the maximum sum available to them unless by reason of exceptional circumstances, such a sum would be unreasonable. This means the RRO is usually 100% of the housing benefit (or universal credit) or rent paid. Where the tenant is applying for an RRO and the landlord has already been convicted of or received a fixed penalty for a failure to licence offence the RRO should be made for such sum as the FTT thinks is reasonable in all the circumstances.

In Wales, an RRO for the maximum sum is only ever available to a local authority seeking an RRO in respect of benefits paid for a failure to obtain a licence and where a conviction has been secured or the Tribunal is satisfied that the offence has been committed. A tenant is only ever entitled to an RRO for such sum as is considered reasonable in all the circumstances.

Where a Tribunal in Wales is making an RRO for such sum as is considered reasonable, then it is required to take account of the following factors:

- The total amount of rent paid during the period that the landlord was committing an offence for failing to have a licence;
- The extent to which the rent was paid by housing benefit and how much was received by the landlord;
- Whether the landlord has been convicted of an offence for failing to have a licence;
- The conduct and financial circumstances of the landlord; and
- The conduct of the occupier.

These factors have been considered in two key UT cases.[31] These are binding on the FTT and therefore provide authoritative guidance on how to determine the level of rent repayment orders.

The UT took the view that the legislation makes clear that the amount to be repaid must be a sum that is "reasonable in the circumstances". Considering the Tribunal's broad discretion, there are a range of factors that should be taken into account when deciding RRO applications. For example, the Tribunal should bear in mind the fact that the landlord is potentially liable for two penalties. First, a fine for the criminal conviction and secondly an RRO. The Tribunal must therefore have regard to the total amount that the landlord will have to pay. The Tribunal should also consider the seriousness of the offence and the culpability of the landlord. A professional landlord deliberately failing to obtain a licence is likely to merit a higher award than the inadvertent oversight of a non-professional landlord.

As the purpose of the RRO is to penalise a landlord profiting from letting an unlicensed property, consideration should be given to the costs of running the property. Where utility bills are incorporated in the rent, these costs should be included in the calculation of the rent received by the landlord but should only be included in the RRO in extreme cases. In *Parker*, the Tribunal also deducted the landlord's mortgage repayments from the RRO.

Importantly, the UT concluded that conduct on part of landlord unrelated to the offence is irrelevant. This is important because tenants often assert poor conduct on behalf of the landlord to justify a higher RRO. In *Parker*, one of the tenants argued that Mr Parker had intimidated and harassed him and had failed to carry out repairs. The UT concluded that these issues did not form part of the underlying offence of failing to have an HMO licence and therefore could not be used to increase the RRO. To do so would be to punish the landlord for matters outside the offence for which he had been convicted.

The final point made by the UT is that there is no presumption that there should be an award of 100% repayment of their rent for the relevant period unless there are good reasons for not doing so. It is wrong to take the maximum award as the starting point and then apply relevant factors to depart from the full amount. An RRO is not really like sentencing in which magistrates proceed from a starting point and then make adjustment to that penalty to arrive at a fine. Instead, the Tribunal must take all the relevant factors into consideration and determine overall what is reasonable in the circumstances.

In England, the legislation on RROs was substantially altered when it was revised to allow for tenants to make an RRO without a preliminary prosecution by the local authority.[32] The amended RRO structure in England states that an RRO must be for no more than 12 months' rent relating to a period during which rent was paid by the tenant and while a relevant offence was being committed by the landlord.[33] The legislation provides that when determining the value of the award, the Tribunal must consider the conduct of both parties, the financial circumstances of the landlord, and whether the landlord has had any prior HMO convictions. Importantly, the requirement of reasonableness found in the legislation before it was amended, and which is still present in Wales, was removed in England.

The relatively new RRO position in England and the way in which the legislation is written – a combination of allowance for judicial discretion and poor drafting that is too vague – means that one has to review a considerable number of Tribunal decisions to find the correct interpretation of some provisions. The Tribunal has considered key questions that go to the heart of the RRO regime such as who can be the subject of an RRO and what considerations must be taken into account when deciding what award to make.

While any landlord – direct or superior – may hold liability for an HMO offence,[34] this does not necessarily mean that both direct and superior landlords may also be the subject to an RRO application. As already explained, the RRO process is one of *repayment* of rent or housing benefit, which is in addition to the civil and criminal penalties a landlord may face for committing an HMO offence.

Rent-to-rent arrangements (where a superior landlord lets a property to an intermediary who in turn sub-lets it to the end occupier(s)) have become increasingly common in the PRS in recent years. In that context where a chain of landlords exists, the question of who may be a Respondent to an RRO application has become the subject of a number of contradictory judicial decisions over a relatively short span of time.

The legislation provides that the Tribunal may make an RRO where a relevant offence has been committed by "a landlord".[35] Thus, the Act does not clearly limit liability for an RRO to a direct landlord, nor does it exclude liability for a superior landlord. Even receipt of rack-rent is not a definitive indicator of whether liability for an RRO is limited to the direct landlord and the UT has accepted that more than one landlord can be receiving rack-rent at the same time.[36]

Rent-to-rent arrangements often involve the property being sub-let as an HMO. While such an arrangement may have its benefits to both parties where the owner of the property receives hassle-free guaranteed monthly income and the rent-to-renter profits from the increased rent charged to sub-tenants, this is a complex commercial setup which may be very problematic in the context of HMOs. Inadequate agreements (usually poorly adapted standard Assured Shorthold Tenancy,

AST, agreements) are commonly used, which do not correctly reflect the nature of the relationship and fail to deal with important matters, e.g. who is responsible to license the property.

Additionally, in a rent-to-rent arrangement, the owner (or other superior landlord) usually hands over full control of the property to the rent-to-renter. This makes it difficult to ascertain whether the property is in fact being run as an HMO when it should not be. Property owners could thus be exposed to a risk of being prosecuted, fined, or having an RRO made against them even where a rent-to-renter is operating an unlicensed HMO without the superior landlord's knowledge or consent.

In *Rakusen*, an inadequate AST agreement was used between a property owner and a tenant in a rent-to-rent arrangement. The tenant then sub-let the property in a manner which made it a licensable HMO under part 2 of the Housing Act 2004.

A few months after the rent-to-rent arrangement was terminated, the sub-tenants made an RRO application against the property owner on the grounds that he had control over or managed an unlicensed HMO. This was challenged on the basis that the Tribunal had no jurisdiction to make an RRO against the property owner as he was not the direct landlord of the sub-tenants. However, both the FTT and then the UT did not accept that argument holding that an RRO could be made against any landlord up the chain, even if there was no direct relationship between the applicant and that person.

Interestingly, the UT stated that interpreting the legislation to allow an RRO to be made against any landlord would best give effect to the objective of the HMO regime, which seeks to discourage the activities of rogue landlords in the residential sector by imposing stringent penalties. The UT was concerned that the alternative interpretation could facilitate a culture where a superior landlord could grant a short-term tenancy to an intermediary for the purpose of avoiding liability, which would render the HMO regime substantially less effective.

The Court of Appeal (CoA) disagreed.[37] The CoA found that the wording of the legislation was clear and unambiguous stating that the natural interpretation was that an RRO can only be made against the immediate landlord of the applicant. Had Parliament truly intended for superior landlords to be subject to RROs, then "it would have been easy enough to make clear and express provision for it, and it has not done so".[38] The CoA relied on the principle of statutory interpretation that "a person should not be penalised except under clear law".[39]

So, while only a direct landlord can be a subject to an RRO, any superior landlord remains exposed to criminal liability[40] and civil penalties. However, as this book is being finalised, the Supreme Court has granted permission to appeal in *Rakusen* so we are yet to hear the last word on this subject.

The above provides a flavour of the difficulties faced by parties trying to navigate the RRO regime and emphasises the challenges faced by Judges in interpreting the law. This theme is continued when one starts to consider the correct value of RRO awards.

In *Ficcara*,[41] the UT considered whether more than one RROs may be made where a landlord has committed more than one relevant offences.[42]

An application for an RRO must be made within 12 months of the offence being committed.[43] The amount of the award is determined by reference to either rent paid by the tenant in the period of 12 months ending with the date of the offence for harassment and unlawful eviction[44]; or during the period when the HMO licensing offence was being committed, not exceeding 12 months.[45]

In *Ficcara*,[46] the UT and the FTT agreed that the Tribunal's powers to determine the award were limited to the 12 months' rent paid by the tenant during the relevant period irrespective of the number of offences committed by the landlord. Emphasis was placed on the use of the word "repayment" in the legislation and the fact that tenants had other means to be compensated for unlawful eviction and harassment (two of the offences committed by the landlord in addition to the HMO licensing offence). However, the committal of several offences could be taken into consideration when determining the value of the award. The purpose of the RRO process was to deter landlords rather than to compensate tenants.

By way of analogy to tenant deposit protection legislation where there is a clear provision in the law for awarding a multiple of the tenant's deposit,[47] the UT would have only awarded an award higher than 12 months' rent where Parliament had set out its intention to allow this in clear words, in accordance with the presumption against doubtful penalisation.[48]

But how should the Tribunal determine the value of an award and what factors should it take into consideration? The legislation provides that when determining the value of the award the tribunal must consider the conduct of both parties, the financial circumstances of the landlord, and whether the landlord has had any prior HMO convictions.[49] There used to be a provision for reasonableness in the legislation but this was repealed in England and is no longer relevant, although it continues to apply in Wales. Nevertheless, it might be assumed that a Tribunal, as a public body, should always act reasonably, and for some time, this approach continued despite the alteration in wording.

However, this approach was somewhat upset when the UT confirmed that the reasonableness of the award was no longer relevant.[50] This meant the landlord's profit was not to be considered anymore and the landlord's expenditure in respect of the property, such as mortgage payments and expenses resulting from meeting obligations to the tenants, was not to be deducted when considering what award to make. At this point, the UT was clear that the starting point for an RRO award was the full rent paid by the tenant during the relevant period.

This was not the last word on the matter however and, largely in order to impose some clarity on the increasingly confusing situation, the President of the UT, Mr Justice Fancourt, in apparent recognition of the importance of the issues being considered became involved and dealt with an RRO appeal himself.[51] The role of the UT President is largely titular and it is unusual for the President to hear appeals. While his view on RROs is technically only binding on the FTT and not on other UTs, the reality is that a decision by the President is very important and will not be ignored. The UT clearly intended to make a point with this decision and streamline the approach taken on RRO awards.

Therefore, the correct approach for assessing RRO awards is now that:

1 The reasonableness of the amount is irrelevant and the landlord's profit does not constitute a limit to the size of an RRO award.
2 The maximum award is not the starting point in the assessment of the amount.
3 The Tribunal should take into consideration the factors set out in s.44(4)(a), any other factors that appear to be relevant, and the context of the purpose underlying Part 4 of the 2016 Act when assessing what award should be made.

To some extent, the UT has turned full circle and brought the position back to what it was before the process started. That is, that there is no specific starting point in setting an appropriate level of the RRO. The Tribunal should take into account the entirety of the facts and then set a penalty based on that consideration and not take an artificial position of starting at a specific proportion of the rent, whether 0%, 50%, or 100%, and then adjusting from there.

Proceeds of Crime

Where an offence has been committed, there are provisions which allow for the recovery of the money earned by the commission of that offence, known as the proceeds of crime.[52] This process is known as confiscation. This is a somewhat misunderstood term. What is being confiscated is the money earned by the unlawful activity and not necessarily all the property owned by an offender, although there can be circumstances where that is a possibility.

However, for money to be confiscated, it must be the proceeds of criminality. That means the person involved must have been convicted of an offence and the money must have been obtained in a criminal manner. In other words, the money earned must result from the criminality. So, for example, rent paid in respect of an unlicensed HMO is not money that can be made subject to a confiscation order because it has not been earned criminally.[53] This is because the obligation to pay rent and to perform all the elements of the tenancy agreement continues whether or not a licence has been obtained or a crime committed.[54] The illegality is the failure to obtain a licence, not the renting of the property or the earning of money from it. In addition, the purpose of the confiscation regime is to "deprive the defendant of the product of his crime or its equivalent, not to operate by way of fine".[55] So, allowing for confiscation where there is already a means to recover rent paid by way of rent repayment order would potentially allow for double recovery and be a fine, which is not the purpose of the confiscation regime.

However, this does not mean that there are no other ways in which a landlord might be subject to confiscation proceedings.

For example, it has been held that operating property in breach of a planning enforcement notice is an offence which allows the proceeds to be confiscated.[56] So, a landlord who operated an HMO in breach of a planning requirement and then further breached an enforcement notice could be prosecuted for that breach and would also then face the possibility of having the proceeds of that offence, the rent earned after the enforcement notice came into effect, confiscated.

In addition, it has been held by junior courts, although this is something that an appeal court may one day disagree on, that breaches of HMO Management Regulations and, in specific cases, rents received when not in possession of a selective licence might be susceptible to confiscation.[57]

Where a property has been used in breach of a selective licence as an HMO and therefore it has been overcrowded, then the excess rent obtained by this overcrowding is the proceeds of a criminal act and will therefore be subject to confiscation. It may also be the case that where an HMO licence has been obtained but the property has had more people in it than permitted by that licence, then that extra rent obtained by overcrowding beyond the specified limit in the licence is also subject to confiscation.

In addition, it has been held that a failure to comply with the HMO Management Regulations or specific licence conditions would give rise to a financial benefit in not doing the works required by the relevant regulation or condition and that this benefit is the proceeds of crime and so subject to confiscation. However, it can be hard to calculate this benefit and the only case to date involved a complex argument between experts as to what benefit had been obtained by the failures. As any work not done is putative, the cost can be difficult to assess.

Notes

1 See *Wandsworth London Borough Council v Rashid* [2009] EWHC 1844 (Admin).
2 *Attorney General's Reference* (No. 1 of 1990) (1990) Times, 21 June.
3 See https://www.cps.gov.uk/publication/code-crown-prosecutors [last accessed 19 February 2019].
4 *R (on the application of Mohamed Lahrie Mohamed & Shehara Lahrie) v Mayor and Burgess of the London Borough of Waltham Forest* [2020] EWHC 1083 (Admin).
5 Under s17, Equality Act 2010.
6 See para 3.5, Ministry of Housing, Communities, and Local Government (2018), *Houses in Multiple Occupation and Residential Property Licensing Reform: Guidance for Local Housing Authorities*, London: Ministry of Housing, Communities, and Local Government [Accessed on 23 March 2022] https://www.gov.uk/government/publications/houses-in-multiple-occupation-and-residential-property-licensing-reform-guidance-for-local-housing-authorities
7 S127(1), Magistrates Courts Act 1980.
8 *Brentford Justices, ex parte Wong* [1981] QB 445.
9 S179, Town and Country Planning Act 1990.
10 *Bow Street Stipendiary Magistrate, ex parte DPP* (1989) 91 Cr App R 283.
11 Sections 72(4) and 95(3), Housing Act 2004.
12 See Newman CJ and Middleton B (October 2010) "Any Excuse for Certainty: English Perspectives on the Defence of 'Reasonable Excuse'", *Journal of Criminal Law* 74, no. 5: 472.
13 *Wandsworth London Borough Council v Sparling* (1987) 152 JP 315.
14 *Aytan & Ors v Moore & Ors* [2022] UKUT 27 (LC).
15 As a result of section 85, Legal Aid Sentencing and Punishment of Offenders Act 2012.
16 Under sections 171D, 179 and 187 of the Town and Country Planning Act 1990.
17 Sections 187B, Town and Country Planning Act 1990.
18 Local Government Association, *Supporting a Thriving Private Rented Sector*, June 2014. Available at https://www.local.gov.uk/sites/default/files/

documents/supporting-thriving-priva-dca.pdf [Accessed on 7 February 2019].
19 Section 143(1), Criminal Justice Act 2003.
20 See https://www.sentencingcouncil.org.uk/overarching-guides/magistrates-court/item/seriousness/ [Accessed on 7 February 2019].
21 See https://www.sentencingcouncil.org.uk/overarching-guides/magistrates-court/item/totality/ [Accessed on 7 February 2019].
22 See https://www.sentencingcouncil.org.uk/overarching-guides/magistrates-court/item/reduction-in-sentence-for-a-guilty-plea-first-hearing-on-or-after-1-june-2017/ [Accessed on 7 February 2019].
23 See https://www.sentencingcouncil.org.uk/overarching-guides/magistrates-court/item/general-guideline-overarching-principles/ [Accessed on 1 September 2022].
24 *Opara v Olasemo* [2020] UKUT 96 (LC) (31 March 2020).
25 *London Borough of Waltham Forest v Marshall* [2020] UKUT 35 (LC).
26 See https://www.gov.uk/government/publications/a-fairer-private-rented-sector/a-fairer-private-rented-sector [Accessed on 1 September 2022].
27 *Sumal & Sons (Properties) Ltd v London Borough of Newham* [2012] EWCA Crim 1840. The Supreme Court refused permission to appeal this case further.
28 In Chapter 4 of the Housing and Planning Act 2016.
29 *Opara v Olasemo* [2020] UKUT 96 (LC).
30 *Camfield & Ors v Uyiekpen & Anor* [2022] UKUT 234 (LC).
31 *Parker v Waller* [2012] UKUT 301 (LC); *Fallon v Wilson & Ors* [2014] UKUT 0300 (LC).
32 Chapter 4, Housing and Planning Act 2016.
33 Section 44, Housing and Planning Act 2016.
34 *Urban Lettings (London) Ltd v Haringey* [2015] UKUT 104 (LC).
35 S.40(a), Housing and Planning Act 2016.
36 *Urban Lettings (London) Ltd v LB Haringey* [2015] UKUT 104 (LC).
37 *Rakusen v Jepson & Ors, Safer Renting Intervenor* [2021] EWCA Civ 1150.
38 [2021] EWCA Civ 1150, para 56.
39 *Bennion on Statutory Interpretation* (7th ed) at 27.1; [2021] EWCA Civ 1150, para 43.
40 *Urban Lettings (London) Ltd v Haringey* [2015] UKUT 104 (LC).
41 *Rosa Ficcara and others v Hannah James* [2021] UKUT 0038 (LC).
42 Housing and Planning Act 2016.
43 Section 41(2)(b), Housing and Planning Act 2016.
44 Section 44(2), Housing and Planning Act 2016.
45 Section 44(2), Housing and Planning Act 2016.
46 [2021] UKUT 0038 (LC).
47 Section 214(4), Housing Act 2004.
48 Bennion on Statutory Interpretation, 7th ed, para 26.4.
49 Section 44, Housing and Planning Act 2016.
50 *Vadamalayan v Stewart* [2020] UKUT 0183 (LC).
51 Williams v Parmar & Ors [2021] UKUT 244 (LC).
52 In Part 2 of the Proceeds of Crime Act 2002.
53 *Sumal & Sons (Properties) Ltd v London Borough of Newham* [2012] EWCA Crim 1840. Permission to Appeal this case further was refused by the Supreme Court.

54 Section 73(3), Housing Act 2004.
55 *Crown Prosecution Service v Jennings* (Appellant) [2008] UKHL 29 @ 13.
56 *R v Del Basso and Another* [2010] EWCA Crim 1119.
57 *London Borough of Brent v Shah and Others*, unreported, 29 January 2018, Harrow Crown Court.

6 HMO Management

All Houses of Multiple Occupation (HMOs), irrespective of licensing status, are required to comply with the Management of Houses in Multiple Occupation (England) Regulations 2006 and their equivalent in Wales,[1] which is exactly the same. Failure to comply with these regulations is an offence which is prosecutable in the magistrates' court and, in England only, can also be dealt with by way of a fixed penalty notice.

Application

It is important to understand that the HMO Management Regulations apply to all s.254 HMOs, irrespective of whether or not they need a licence. That means that HMOs which do not need licences still have a set of standards which apply to them and for which a local authority can bring a criminal prosecution if they are not adhered to. A similar set of regulations have been passed for s.257 HMOs[2] which set out obligations which primarily relate to the common parts of these properties.

Requirements

The regulations require a number of things, including that:

- the property is clean at the start of the tenancy;
- the property and the immediate area, including front and rear gardens, are kept tidy;
- the property is kept in repair;
- the property is fire safe;
- steps are taken to protect the occupiers from injury;
- a notice is placed in the property giving the contact details of the manager; and
- the fixed electrical installation is tested every five years.

Although the regulations use the word manager or management, these obligations fall specifically on the person managing the HMO, which will usually be the person collecting the rent. This occasionally causes issues as local authorities again often read the wording of the regulations plainly and consider that a person described as the manager is liable for complying with the management regulations.

Licence Conditions

As well as the more general obligations imposed in relation to HMOs by the HMO Management Regulations, any licensable property will have a range of conditions associated with it. Many of these will also either replicate the HMO Management Regulations or will impose very similar obligations in relation to selectively licensed properties which do not have regulations like the HMO Management Regulations which apply to them. Failure to comply with licence conditions is also an offence. Interestingly, it is the one offence that is not committed by a person managing or person having control but can, in theory, be committed by someone else. The person liable for a breach of a licence condition is the licence holder. In the vast majority of cases that person will also be collecting or receiving the rent or would have been entitled to receive it and so they will also be a person managing or having control. But, at least in principle, they could be a different person.

Specific Obligations

The various requirements in the HMO Management Regulations or in licences often cause difficulty so I outline some of the issues here.

Cleaning

It is not acceptable to simply rely on the occupiers to keep things clean, even in a property where the occupiers are in control of the entire space. This is even more the case where there are common areas in a shared bedsit-type property that the landlord has effective control of and can enter whenever he or she wishes. It is well worth considering a professional cleaner for the common areas and adding the cost to the rent if the tenants cannot be relied on to keep the premises clean as the landlord will be held liable for this. If a professional cleaner is being used, then this should be a genuine commercial arrangement and not an 'off the books' payment to one of the tenants. These are probably unlawful as there will be no following of the usual employment formalities and there is a risk that the tenant will not in fact do the work.

Repair and Safety

There are several obligations regarding repairs and safety. Many landlords think that if they have not had the repairs reported to them, then they cannot be liable and then complain about prosecutions started by local authorities without, as they see it, giving them a chance to make repairs. This entirely misses the point. When talking about disrepair under s.11, Landlord & Tenant Act 1985, it is correct to say that landlords are not required to carry out repairs until they are put on reasonable notice. However, this is an addition by the courts and is not written into the legislation. The requirement under the HMO Management Regulations and under any licence is one to ensure that premises are safe and that they are in repair. This is different, and there is absolutely no requirement for a landlord to get notice first. It is up to the landlord to have an effective and regular inspection regime in place to identify the need of repair and execute it without relying on notice from occupiers or local authority officers.

The same applies to safety. This is largely focused on the more serious safety breaches and particularly on the possibility of falls. So, landlords are expected to ensure that the property has suitable protective bannisters and railings around stairs and balconies, restrictors on windows so that they cannot easily be opened too far, and ensure that flat roofs are safe or that they are inaccessible.

Fire

There are substantial requirements in regard to fire safety. The guidance produced by LACORS is still the main guide used by local authorities when considering safety standards in residential premises.[3] This is frustrating as LACORS no longer exists and the guidance is quite hard to locate. It is also somewhat outdated having been published nearly ten years ago and so does not take into account substantial changes in property use and fire safety technology since that date. A clarification document was issued in March 2009 and the main guidance should be read with regard to the clarification document.[4]

There is no legal requirement to produce a fire risk assessment for a domestic residential property as the Regulatory Reform (Fire Safety) Order[5] (RRFSO) will not apply. However, where a property is an HMO made up of bedsits with shared use of common areas and cooking and washing facilities, the RRFSO will apply and a fire risk assessment should be produced and recorded in writing.[6] Even in properties where a risk assessment is not required, it is a very good idea to have one as it demonstrates clearly that the landlord has considered fire safety issues seriously and is committed to dealing with them. As a general guide, the minimum standards that should be aimed at is for there to be:

- Mains powered, battery-backed, interlinked smoke, and heat detectors on at least each floor and ideally in every major room;
- Automatic closers on key doors such as bedrooms;
- Intumescent smoke seals on key doors and ideally cold smoke seals;
- All doors to be at least 30 minutes fire-proof; and
- Walls to be robust enough that they can provide 30 minutes fire proofing.

More complex or larger properties may need more than this, including specialised lighting, signage, and more substantial safety measures.

When considering fire safety, many landlords forget about the need to manage the occupiers. There is little point in having a protected fire escape route if it is blocked by bicycles or other paraphernalia from the occupiers. Equally, landlords must take care not to store or abandon potentially flammable decorating materials such as paint in high-risk areas such as under-stairs cupboards or meter cupboards. It is important that landlords are ensuring when they inspect that they consider fire safety issues and take clear action in respect of them to correct tenant behaviour or to resolve issues and to charge them back to the tenant at a later date.

Notices

This is a simple obligation but one that is often ignored. There is an absolute requirement to ensure that the tenants have the manager's name, address, and telephone number and also that that same information is displayed prominently in the property. This can be a problem as notices in the property are often torn down or defaced. Some magistrates have accepted that the provision of this information to the tenants by giving each of them a specific personal notice and asking them to sign to confirm that they had received it is acceptable. However, this is high-risk and a notice with some photographic evidence of its placement would be sensible. There is little excuse for failing to comply with this obligation, and it is routinely ignored by landlords, especially in unlicensed property.

Electrics

There is a general requirement that electrics must be safe as well as a more specific requirement that the landlord must obtain a certificate that the fixed wiring is safe every five years. The requirement is to have a certificate for the entire installation and so the practice that has been adopted by some landlords of having a bit checked each year will only be legal if they initially have a full check and then continue annually from that starting point so that at all times all of the installation is covered by a certificate that is less than five years old.

It is important to consider the electrical installation in the more practical sense of how it will be used as well. Most Environmental Health Officers will consider that an installation that has plug sockets near sinks or where the tenant has insufficient plug sockets which compels the tenant to use numerous extenders and adaptors and to overload sockets is unsafe. They are likely to be correct in this. Landlords should be aiming for multiple sockets in rooms that are being lived in, especially in HMO property and should consider more modern sockets that have USB ports for charging electronic devices as well.

Active Management

The main problem for most managers of HMOs and other licensed properties is that they adopt the same management methods as they would for more traditional lettings. This operates as a largely passive form of management in that there is a reliance on tenants reporting problems which are then resolved. This is not an option under the HMO Management Regulations or in relation to most licensed properties where a far more pro-active approach is required. This is because a local authority can inspect the property at any time and will treat any defects they find there as a breach of the landlord's obligations or licensing conditions.

In some properties, this will be easier than others depending on how they have been let and the style of property. It would be advisable to risk assess tenants at an early stage and then decide on the appropriate schedule of inspections.

Most local authority officers will tend towards the view that the industry norm of quarterly inspections is not sufficient, especially in HMO property and certainly in bedsit style HMOs where the landlord has an unfettered right to access the common areas whenever he wishes.

It appears that the Upper Tribunal agrees with the tough approach. It has found that a civil penalty issues in respect of a landlord who breached the fire safety requirement was acceptable, even though the landlord had a robust inspection and resolution system in place. It was being penalised for matters that it had first become aware of on the Council's inspection. The UT also did not accept there was a reasonable excuse. Ultimately, the position is that the management regulations create a strict liability in respect of any failure.[7]

Risk Assessment

The most practical way forward is to risk assess tenants and properties in writing and to use this assessment to set out an inspection schedule. This should be done based on the nature of the occupation and an assessment of the tenants' behaviour. Factors that would increase the frequency of inspections might be the age of the tenants, their social demographic (students, for example, might need more inspections), the number of occupiers, the type of tenancy and property (bedsit as against shared house), and the level of sharing of facilities. Any risk assessment should be based in part on a visit conducted in the first few weeks to allow practical observation of cleanliness, wear and tear caused by the occupiers, and any fire safety or other hazards that might need addressing. Based on this, a full risk assessment can then be completed in writing along with a plan on regularity of inspection.

Records

In addition to inspections, there is a need to ensure proper records are kept of inspections and actions. Most landlords and agents do not keep sufficient records of inspections, any actions taken after inspections and of notifications to tenants of breaches found during those inspections, and what has been done to follow up on issues. Many landlords and agents keep no records at all. If there are no records, it is very hard to show whether a defect found by a local authority officer is a one-off that has just occurred or is in fact symptomatic of poor management. Ideally, a checklist showing inspections should be kept with each part of the property shown, consideration given to key items, clear evidence that action was taken, and evidence that there was a follow up to ensure the problem was resolved.

Notes

1 The Management of Houses in Multiple Occupation (Wales) Regulations 2006.
2 See the Licensing and Management of Houses in Multiple Occupation (Additional Provisions) (England) Regulations 2007 and the Licensing and Management of Houses in Multiple Occupation (Additional Provisions) (Wales) Regulations 2007.
3 See LACORS (July 2008) *Housing – Fire Safety: Guidance on Fire Safety Provisions for Certain Types of Existing Housing.* London: LACORS. [Accessed on August 2017]. Available at: https://www.rla.org.uk/docs/LACORSFSguideApril62009.PDF
4 LACORS, *Clarification Paper Issues by The Housing Fire Safety Steering Group,* March 2009. [Accessed on August 2017]. Available at: https://www.rla.org.uk/docs/clarificationlacors310309.pdf
5 Regulatory Reform (Fire Safety) Order 2005/1541.
6 Regulatory Reform (Fire Safety) Order 2005/1541, Article 9.
7 *Adil Catering Ltd v The City Of Westminster Council* [2022] UKUT 238.

7 The Tribunals

Almost all disputes relating to HMO property are heard by the First-tier Tribunal (Property Chamber)[1] (FTT) or the Welsh Residential Property Tribunal in Wales.[2]

Procedural Rules

The Tribunals have their own specific procedural rules. This is complicated by the fact that while there are several FTTs in England, the Property Chamber has its own distinct set of rules that apply. The Upper Tribunal (UT) also has its own distinct set of procedural rules.

The Tribunal Procedure (First-tier Tribunal) (Property Chamber) Rules 2013[3] set out the procedural rules for the FTT in England. The rules include provisions similar to those prescribed by the Civil Procedure Rules for the civil courts, such as:

- An overriding objective for the Tribunal to deal with cases fairly and justly.[4]
- General case management powers empowering the Tribunal to give directions to manage the conduct or disposal of a case.[5]
- Powers to deal with parties' failure to comply with the Tribunal's rules,[6] including the power to refer a failure to the UT and ask the UT to exercise its powers to compel the giving of evidence or production of documents.[7]
- The power to strike out a case for reasons, including failure to comply with directions or for want of jurisdiction.[8]
- Powers to add, substitute, or remove a party[9] and to charge fees.[10]
- A power to make costs orders,[11] which is exercised rarely and is examined below in reference to the relevant case law.
- How to calculate time[12]; file and serve documents[13]; deal with disclosure,[14] expert evidence,[15] and summon witnesses.[16]

DOI: 10.1201/9781003297796-7

Furthermore, the Property Tribunal rules include special powers related to the specialist nature of the Tribunal's work, such as a power to inspect the subject property.[17]

Appeals and Time Limits

An appeal of a decision (e.g. against a local authority notice, order or licence) must be submitted to the FTT within 28 days of the applicant being notified of the relevant decision.[18]

To commence proceedings in the Tribunal, the applicant must send or deliver to the Tribunal a notice of application.[19] After processing the application notice, the Tribunal will provide copies to the respondent and any interested persons[20] and directions will be given for a response[21] and sometimes a further reply by the applicant.

Transfers Up

There is an important power to transfer cases upwards from the FTT to the UT.[22] This can be at the request of the parties or by the FTT of its own volition. Both have occurred but it is a power used sparingly although increasingly so compared to when it was initially introduced. Its main purpose is to allow the UT to deal with complex points of law and to provide a more definitive position on specific areas of uncertainty in the law so that further FTT decisions will be bound to follow them. This increases consistency in FTT decision-making and helps to settle areas in which the law is less clear. The power may also be used where a case involves a large financial sum. The decision on whether a case should be transferred up is made by the President of the Property Chamber, with the agreement of the President of the Lands Chamber.

Paper and Oral Determination

The FTT will usually hold a hearing to decide a case,[23] however, it may dispose of proceedings without a hearing where:

- The parties have consented to a paper determination (consent may be deemed where a party has failed to object to a notice that the FTT intends to dispose of the need for a hearing)[24]; or
- The Tribunal is exercising its power to strike out a case.[25]

Where the Tribunal is making an oral determination, hearings will usually be public unless the Tribunal directs otherwise.[26] A hearing may also sometimes proceed in the absence of a party where they have been notified of the hearing and the Tribunal considers it in the interest of justice to proceed with the hearing.[27]

Parties may request the Tribunal to approve a consent order disposing of the proceedings, but the Tribunal will only do so if it considers it appropriate.[28]

The Tribunal may give a decision orally at a hearing[29] but it must provide a written notice of the decision to the parties, accompanied with reasons for the decision and information of any right to appeal.[30]

Re-Hearings and Expertise

It is important to recognise that the role of the FTT is not as a point of appeal. Therefore, it is not there to simply agree with or overturn a local authority decision. This is a point frequently misunderstood by everyone involved, sometimes including the FTT! In fact, all hearings in the FTT under the Housing Act 2004 proceed by way of re-hearing. This means that the FTT is there to make the decision again itself using its own expertise. This does not mean that they ignore the previous local authority decision, but they are certainly not bound by it. Nor are they bound by the facts found by the local authority in relation to the property. In general, the FTT will steer a middle course, using the findings and views of the local authority but also using its own expertise to depart from that when appropriate.

The power of the FTT to depart from the views of local authorities and from local authority internal guidance has proved controversial. There has been an ongoing debate as regards the level of respect that the FTT must give to a local authority decision and to local authority guidance. A number of local authorities have sought to argue that the FTT should accept local authority guidance unquestioningly and should only decide if the local authority has complied with that guidance or not.

The UT has rejected that line of argument. The FTT is an expert tribunal and must be given credit for this. It differs in that regard from a court which only truly has expertise in the law and its exercise and must rely on external experts when judging the suitability of accommodation, for example. The UT has specifically drawn a distinction between areas that are very much within the expertise of the FTT, such as the provision of space and standards in an HMO, in which the FTT should consider local authority guidance but may depart from it freely,[31] and scenarios where the FTT has a lesser degree of expertise, such as the proper application of civil penalties, in which it should be more cautious about departing from local authority guidance and give good reasons when it does so.[32]

Appeals from the FTT

Initially, very few cases were appealed from the FTT. The FTT was very reluctant to give permission for appeals and the UT also was reluctant to give permission.

Before deciding whether to grant permission to appeal, the Tribunal must consider whether to review its decision[33] and may decide to review if it is satisfied that a ground of appeal is likely to be successful.[34] If permission is refused by the FTT, a party may make a further application for permission to the UT.[35] A stay of the implementation of the decision (or a part of the decision) may also be applied for at the same time as permission for appeal is sought.[36]

Beyond the UT

The UT is effectively an equivalent to the High Court. Indeed, many of its judges are also High Court judges as well. Therefore, the appeal route from the UT is directly to the Court of Appeal. As normal in these matters, permission is needed from the UT itself or from the Court of Appeal. Beyond that, appeal is to the Supreme Court. Perhaps a little surprisingly, this has happened.[37]

Costs

Generally, the FTT is a cost neutral venue. However, the tribunal may make an order for costs where a party "has acted unreasonably in bringing, defending or conducting proceedings".[38] Applications for costs in the Tribunal are rare, as it is well-known that the threshold for a successful application is very high. More commonly used is the provision allowing parties to seek a reimbursement from the other side for the Tribunal fees paid.

The wording of Rule 13 in respect of costs applications is vague and the interpretation of what constitutes unreasonableness has been the subject of dispute in the FTT. However, the UT has considered the issue of costs in relation to service charge disputes.

However, the UT has made clear that the bar is high and

> for a lay person to be unfamiliar with the substantive law or with tribunal procedure, to fail properly to appreciate the strengths or

weaknesses of their own or their opponent's case, to lack skill in presentation, or to perform poorly in the tribunal room, should not be treated as unreasonable.

Conduct will need to be "vexatious, and designed to harass the other side rather than advance the resolution of the case".[39] This is not merely a matter of who has won or lost but rather the manner in which the case is pursued. It is entirely possible to be the winning party in a case and still have a costs order made against you.

Tribunals have generally shown themselves to be disinterested in conduct that has occurred before proceedings are issued before them.[40] Therefore, a mistake by a local authority is not likely to lead to a costs order being made against it. However, where that mistake means that the local authority is inevitably going to lose the proceedings, then carrying on with them may well be unreasonable conduct which might lead to a costs order being made.

Simply seeking to withdraw may not provide and escape. The FTT has to be asked to withdraw proceedings, it is not something that can just be done. In some cases, it has refused permission for withdrawal, often at the behest of the other party, and then dismissed the proceedings with a costs award or permitted a judgement to be made.[41]

Notes

1 https://www.gov.uk/courts-tribunals/first-tier-tribunal-property-chamber.
2 http://rpt.gov.wales/.
3 The Tribunal Procedure (First-tier Tribunal) (Property Chamber) Rules 2013/1169.
4 The Tribunal Procedure (First-tier Tribunal) (Property Chamber) Rules 2013/1169, Rule 3.
5 The Tribunal Procedure (First-tier Tribunal) (Property Chamber) Rules 2013/1169, Rule 6.
6 The Tribunal Procedure (First-tier Tribunal) (Property Chamber) Rules 2013/1169. Rule 8.
7 Tribunals, Courts and Enforcement Act 2007, Section 25.
8 The Tribunal Procedure (First-tier Tribunal) (Property Chamber) Rules 2013/1169, Rule 9.
9 The Tribunal Procedure (First-tier Tribunal) (Property Chamber) Rules 2013/1169, Rule 10.
10 The Tribunal Procedure (First-tier Tribunal) (Property Chamber) Rules 2013/1169, Rule 11.
11 The Tribunal Procedure (First-tier Tribunal) (Property Chamber) Rules 2013/1169, Rule 13.
12 The Tribunal Procedure (First-tier Tribunal) (Property Chamber) Rules 2013/1169, Rule 15.

13 The Tribunal Procedure (First-tier Tribunal) (Property Chamber) Rules 2013/1169, Rule 16.
14 The Tribunal Procedure (First-tier Tribunal) (Property Chamber) Rules 2013/1169, Rule 18.
15 The Tribunal Procedure (First-tier Tribunal) (Property Chamber) Rules 2013/1169, Rule 19.
16 The Tribunal Procedure (First-tier Tribunal) (Property Chamber) Rules 2013/1169, Rule 20.
17 The Tribunal Procedure (First-tier Tribunal) (Property Chamber) Rules 2013/1169, Rule 21.
18 The Tribunal Procedure (First-tier Tribunal) (Property Chamber) Rules 2013/1169, Rule 27.
19 The Tribunal Procedure (First-tier Tribunal) (Property Chamber) Rules 2013/1169, Rule 26.
20 The Tribunal Procedure (First-tier Tribunal) (Property Chamber) Rules 2013/1169, Rule 29.
21 The Tribunal Procedure (First-tier Tribunal) (Property Chamber) Rules 2013/1169, Rule 30.
22 The Tribunal Procedure (First-tier Tribunal) (Property Chamber) Rules 2013/1169, Rule 25.
23 The Tribunal Procedure (First-tier Tribunal) (Property Chamber) Rules 2013/1169, Rule 31(1).
24 The Tribunal Procedure (First-tier Tribunal) (Property Chamber) Rules 2013/1169, Rule 31(2) & (3).
25 The Tribunal Procedure (First-tier Tribunal) (Property Chamber) Rules 2013/1169, Rule 9 & 31(4).
26 The Tribunal Procedure (First-tier Tribunal) (Property Chamber) Rules 2013/1169, Rule 33.
27 The Tribunal Procedure (First-tier Tribunal) (Property Chamber) Rules 2013/1169, Rule 34.
28 The Tribunal Procedure (First-tier Tribunal) (Property Chamber) Rules 2013/1169, Rule 35.
29 The Tribunal Procedure (First-tier Tribunal) (Property Chamber) Rules 2013/1169, Rule 36(1).
30 The Tribunal Procedure (First-tier Tribunal) (Property Chamber) Rules 2013/1169, Rule 36(2).
31 *Clark v Manchester City Council* [2015] UKUT 129 (LC).
32 *London Borough of Waltham Forest v Marshall* [2020] UKUT 35 (LC).
33 The Tribunal Procedure (First-tier Tribunal) (Property Chamber) Rules 2013/1169, Rule 53.
34 The Tribunal Procedure (First-tier Tribunal) (Property Chamber) Rules 2013/1169, Rule 55.
35 The Tribunal Procedure (First-tier Tribunal) (Property Chamber) Rules 2013/1169, Rule 53.
36 The Tribunal Procedure (First-tier Tribunal) (Property Chamber) Rules 2013/1169, Rule 54.
37 In Nottingham City Council v Parr & Anor [2018] UKSC 51.
38 The Tribunal Procedure (First-tier Tribunal) (Property Chamber) Rules 2013/1169, Rule 13.

39 *Willow Court Management Company (1985) Ltd v Alexander* [2016] UKUT 290 (LC) (21 June 2016).
40 *Distinctive Care Ltd v Revenue and Customs* [2019] EWCA Civ 1010.
41 *Albert House Property Finance PCC Ltd and another v HMRC* [2020] UKUT 373.

8 Conclusion

The Housing Act 2004 has fundamentally altered the management and operation of multiple occupied property. It has also proved surprisingly enduring in this area with most changes being used to improve enforcement rather than watering it down or making substantial changes. Its effect on the market has been substantial and, if a measure of success is to alter market practice, then it must be said to have succeeded.

At the same time, it could be argued to have failed. There has been no obvious improvement in the standards of property in the Private Rental Sector. Licensing has not assisted here and there is evidence that enforcement activity in areas which have property licensing is actually lower than in areas that do not.[1] From that viewpoint, licensing has become an exercise in bureaucracy with little practical benefit on the ground.

A large part of the problem has been the complexity of the system, coupled with ongoing tinkering by central government. This has left local authorities constantly striving to catch up with a moving playing field. This difficulty has been exacerbated by the very tight funding position that local authorities have been placed, which limits the resources they have available to define high-quality policies and processes for effective licensing and enforcement. It has also led to a loss of trained and experienced Environmental Health Professionals in the sector as many have chosen to leave rather than continue in such a poor working environment and there has been little ability, or effort made, to attract new blood into the sector. At the same time, the "no costs" position in the First-tier Tribunal has limited the ability of landlords to push back against local authority decisions as the cost of hiring legal assistance often outweighs the potential gain. This has left a number of aspects of the law relatively uncertain and has meant that the 2004 Act is only starting to reach maturity some 20 years after its initial introduction.

At the same time, local authorities must take some of the blame in terms of the manner in which they have operated the licensing regimes. The 2004 Act was clearly intended to move towards a far more risk-based approach to regulatory decision making. The rating structure introduced by the Housing Health and Safety Rating System is the clearest example of this. Many local authorities have, sadly, not operated their schemes in this way and have tended to treat HMO licensing as a box-ticking exercise with little consideration of the actual risks and the importance of balancing risk with availability of lower cost living space. The very inflexible and unimaginative approach adopted by some local authorities to issues of room size is a good example of this.

Whatever view is taken, the law on property licensing is complex and, as local authorities increasingly come to understand and utilise it, is becoming more so. While this book aims to provide a guide to some of its features, it is true to say that there remain considerable uncertainties in the legislation and much that the senior courts are likely to have to consider. It is certain that there will be continued development for some time to come.

Note

1 Wood J and Watkin S, *The Enforcement Lottery: Local Authority Inspections and Notices*. (Manchester: National Residential Landlords Association, 2022). https://www.nrla.org.uk/research/special-reports/enforcement-HHSRS-inspections-notices.

Index

Note: *Italic* page numbers refer to figures and page numbers followed by "n" denote endnotes.

additional HMO licensing 30; sections 59 and 83 for 76n21; setting up of 31–32
Adult Placement Schemes (England) Regulations 2004 for England 24n21
Adult Placement Schemes (Wales) Regulations 2004 for Wales 24n21
anti-social behaviour 53, 54, 80
Assured Shorthold Tenancy (AST) agreement 68, 97
asylum seekers 10–11

breaches 88, 93, 100
Building Regulations 1991 18, 19, 23

Care Quality Commission (CQC) 21
Chartered Institute of Environmental Health 5
civil penalties 91, 113
Civil Procedure Rules 111
cleaning 105–106
Code for Crown Prosecutors 83–84
codes of practice 74–75
Commons Library Briefing paper (2019) 1
companies 62–63
confiscation 100–101
consultations: nature of 32–34; notification of licensing designation 34–37; review and revocation of licensing designations 37–39
converted building test 7–8
Court of Appeal (CoA) 49, 55, 97, 114
CQC *see* Care Quality Commission (CQC)
criminality 100
Criminal Law Act 1977 93
Crown Prosecution Service 83
culpability 89, 91, 95

data protection 42–43
defences 86–87
Disclosure and Barring Service 52
dishonesty offence 49–50
disputes 22
domestic staff, household 16
duly made 42
duty to licence 39–40

educational institutions 20
EEA nationals *see* European Economic Area (EEA) nationals
electrics 108
England: civil penalties in 91; fitness tests 50–51; guidance in 85–86; legislation on RROs 95–96; licence application form 40; local authorities in 51, 54; mandatory licensing in 29; national standards, regulations for 47; rent repayment

orders 93–94; revoking licence in 68; room size standards for 56; selective licensing scheme in 31, 32; time dealing with licence application 44, 45
Environmental Health Officers 108
Environmental Health Professionals 118
European Court of Justice (ECJ) 43
European Economic Area (EEA) nationals 10
European Union (EU) 10, 45
EU Services Directive 45
evidential test 84

Ficcara case 98
final orders 81
fines 88–91
fire 106–107
First-tier Tribunal (FTT) 3, 4, 12, 43, 54, 58, 68, 72, 81, 118; appeals from 112, 114; civil penalty 91; costs 114–115; paper and oral determination 112–113; procedural rules 111–115; re-hearings and expertise 113–114; rent repayment orders 93, 94, 98, 99; temporary exemption notices 65; time limits 112; transfers up 112
fitness 48–52
FTT *see* First-tier Tribunal (FTT)
Full Code Test 84

General Data Protection Regulations (GDPR) 42
Generation Rent 1
guilty pleas 89, 90

harm 89, 91
Health and Social Care Act (2008) 21
HHSRS *see* Housing Health and Safety Rating System (HHSRS)
High Court 27, 33, 40, 43, 44, 62, 114; in Northern Ireland 54
HMO *see* Houses of Multiple Occupation (HMO)
HMO declarations, s.254 tests interpretion 12–14, *13*
HMO Management Regulations 100, 101; active management 108–109; application 104; licence conditions 105; obligations 105–108; records 109; requirements 104–105; risk assessment 109
household 14–16
Houses in Multiple Occupation (Certain Converted Blocks of Flats) (Modifications to the Housing Act 2004 and Transitional Provisions for section 257 HMOs) (England) Regulations 2007: regulations 2 and 5 in 77n61
Houses in Multiple Occupation (Certain Converted Blocks of Flats) (Modifications to the Housing Act 2004 and Transitional Provisions for section 257 HMOs) (Wales) Regulations 2007: regulations 2 and 5 in 77n61
Houses in Multiple Occupation (Specified Educational Establishments) (England) Regulations 2013 24n30
Houses in Multiple Occupation (Specified Educational Establishments) (Wales) Regulations 2006 24n30
Houses of Multiple Occupation (HMO) 1–3, 5, 26; additional licensing 30; disputes 22; examples of different scenarios 22–24; exemptions 19–21; liability for 58–61; mandatory licensing 28–30; potential for change 21–22; potential offences associated with 82; property standards 46–47; selective licensing schemes 30–31
Housing Act (1988) 92
Housing Act (2004) 1–3, 5, 34, 38, 45, 49, 53–56, 58, 62, 93, 97, 113, 118, 119; fines 88; s.254 tests 6–17; s.257 tests 17–19
Housing and Planning Act (2016) 2
Housing Health and Safety Rating System (HHSRS) 55–57, 93, 119

Immigration and Social Security Co-ordination (EU Withdrawal) Act 2020: regulation 67 of 24n11
impartiality 83

independence 83–84
individual local housing 28, 30
interim orders 81

LACORS 106
landlords 60; appeal civil penalty 91; avoid licensing 29; grace period for 39–40, 56; judicially review licensing designation 34; mass licensing in Wales 32; operating unlicensed properties 92; publication of personal data 43; reasonable excuse 37; records 109; responsibility to comply with law 85; temporary exemption notices 63
Landlord & Tenant Act 1985 106
Land Registry 43, 66
legislation: complexity of 2, 22; definition of household 14–15; general tenor of 57; licensing scheme 33; local authorities deal with 2; in Northern Ireland 54; reasonable excuse defence 87; relating to Health and Safety 62; relation to council tax and planning 2; repayment in 98; on RROs 95; secondary 14, 17; for single application fee 44
licence application form 40, *41*; costs of 43–44; data protection and licence applications 42–43; duly made 42; time limit 44–46
licence conditions 52–54; and HHSRS 56–57; HMO Management Regulations 105; room sizes 55–56; in selective licences 54–55
licence standards: fitness and suitability 48–52; HMO property standards 46–47; local standards 47–48; management and suitability 48
licensable property, management take over 80–81
licensees 17
licensing 118, 119; additional HMO licensing 30; applicability of 39–40; codes of practice 74–75; grant and refusal process 57–58; liability HMOs 57–58; licensable property and alternatives 57–58; mandatory HMO licensing 28–30; register of 72–74; revocation of 65–69, *66*; selective licensing schemes 30–31; termination of 65–69, *66*; variation in 69–72, *70*
Licensing and Management of Houses in Multiple Occupation and Other Houses (Miscellaneous Provisions) (England) Regulations 2006: regulation 3 in 24n18; regulation 4 of 24n20; regulation 5 of 24n10; regulation 9 of 76n22; regulation 10 of 76n29; regulation 13 of 79n137; Schedule 1 of 24n35
Licensing and Management of Houses in Multiple Occupation and Other Houses (Miscellaneous Provisions) (Wales) Regulations 2006: regulation 3 in 24n18; regulation 4 of 24n20; regulation 5 of 24n10; regulation 9 of 76n22; regulation 10 of 76n29; regulation 11 of 79n137; Schedule 1 of 24n35
licensing designation 32; judicial review of 34; notification of 34–37; review and revocation of 37–39
listening exercise 33
Local Government Association 88
local standards, licence 47–48

magistrates courts 82, 86, 88, 91, 92
management order 80–81
mandatory HMO licensing 28–30
mens rea 84–85
migrant workers 10, 23
mitigation 87–88

non-EEA nationals 10, 23
Northern Ireland: High Court in 54; HMO licence applications 45; legislation in 54
notices 107–108
numbers: of occupiers 17, 20–21, 28, 47, 52, 56, 57; of offences 84, 98; of PRS properties 55; of storeys 26–29

obligations 105; cleaning 105–106; electrics 108; fire 106–107; notices 107–108; repair and safety 106
offences: associated with HMO 82; mitigation 50; rehabilitation 50; relating to planning 86; relevance 49–50; seriousness of 50, 89
'one size fits all' approach 53
owner-occupied properties 18

parental relationships 15
Parker case 95
penalties: civil 91; rent repayment orders 92–99
person having control 59–62
person managing 59–62
Private Rental Sector (PRS) 1, 55, 118
proceeds of crime 99–101
Proceeds of Crime Act 92
property: active management 108–109; dispute 22; freeholder of 61; large family home 22–23; large house 23; managed/controlled 20; management of 61; old house 23; overcrowding 52; in Private Rental Sector 55; purpose-built block of flats 23; rack-rent 60; regulated under alternative regimes 21; risk assessment 109; sale of 65–66; standards of 118; storeys 26–28; two-storey house 22; unlicensed 46, 92, 95, 108
Property Tribunal rules 112
prosecutions 82; Full Code Test 84; room sizes and public interest 85–86; state of affairs offences and *mens rea* 84–85; time limits 86
prosecutors 82–83, 85, 86
Protection from Eviction Act 1977 93
Provision of Services Regulations (2009) 51
PRS *see* Private Rental Sector (PRS)
public bodies 20
public interest, prosecutions 85–86
public interest test 84, 85

rack-rent 60, 61; defined as 59
Rakusen case 97, 98
reasonable excuse defence 87, 88

reasonable suitability 57
records 109
register of licences 72–74
registration gap 65
Regulatory Reform (Fire Safety) Order (RRFSO) 106–107
religious communities 20
rent repayment orders (RROs) 4, 42, 46, 87, 92–99; direct and superior landlords 96; England, legislation on 95–96; *Ficcara* case 98; FTT 93, 94, 98, 99; *Parker* case 95; *Rakusen* case 97, 98; rent-to-rent arrangements 96–97; tenants 93–94; UT 94–95, 97–99
rent-to-rent arrangements 96–97
repair 106
Residential Property Tribunal (RPT) 3, 4
review, of licensing designations 37–39
revocation: of licensing 65–69, 66; of licensing designations 37–39
revocation notification 37–38
Right to Rent legislation 51
risk assessment 107, 109
Rogue Landlord and Agent database 51
room sizes: licence conditions 55–56; prosecutions 85–86
RPT *see* Residential Property Tribunal (RPT)
RRFSO *see* Regulatory Reform (Fire Safety) Order (RRFSO)
RROs *see* rent repayment orders (RROs)

safety 106
seasonal worker 10
secondary legislation 14, 16, 17
Sedley Criteria 32, 33
selective licensing 30–31, 37, 69; breach of 100; licence conditions in 54–55; sections 59 and 83 for 76n21; setting up of 31–32
selective licensing-related offence 87
self-contained flat test 7, 27
Sentencing Council 89, 90
sole use 11–12
standard test, s.254 6–7

state of affairs offences 84–85
s.254 tests 6, 30, 38, 104; converted building test 7–8; interpretation of (*see* s.254 tests interpretion); self-contained flat test 7; standard test 6–7; summary position for 21
s.254 tests interpretation: buildings and parts of buildings 9; HMO declarations 12–14, *13*; household 14–16; only/main residence 9–11; sole use 11–12; tenants, licensees, and numbers 17; units of living accommodation 9
s.257 tests 17–19, 30, 38, 47, 104
storeys 26–28
suitability, licence standards: fitness and 48–52; management and 48
Supreme Court 54, 98, 114

temporary exemption notices (TEN) 63–65, *64,* 73, 87
tenants 17; cleaning 105–106; electrics 108; fire 106–107; notices 107–108; rent repayment orders 93–94; repair and safety 106; risk assessment 109; variation of HMO licence 69
termination, of licensing 65–69, *66*
time limit: First-tier Tribunal 112; licence application form 44–46; prosecutions 86; Upper Tribunal 112
Tribunal Procedure (First-tier Tribunal) (Property Chamber) Rules 2013 111–112; appeals and time limits 112; appeals from FTT 114; beyond UT 114; costs 114–115; paper and oral determination 112–113; re-hearings and expertise 113–114; transfers up 112
two-stage fee 44

units of living accommodation 9
unlicensed property 46, 92, 95, 108
Upper Tribunal (UT) 3, 4, 11–12, 19, 47, 61, 65, 66, 87; active management 109; appeals and time limits 112; beyond 114; civil penalty 91; costs 114–115; paper and oral determination 112–113; procedural rules 111–115; re-hearings and expertise 113–114; rent repayment orders 94–95, 97–99; transfers up 112

Wales: guidance in 86; local authorities in 54; mandatory licensing in 29; mass landlord licensing in 32; national standards, regulations for 47; rent repayment orders 93–94; selective licensing scheme in 31, 32; time dealing with licence application 44, 45
Welsh Order 26–28
Welsh Residential Property Tribunal (RPT) 53, 58, 69, 72

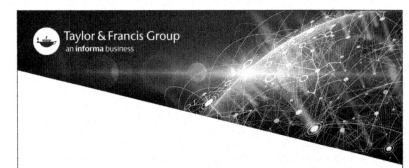